International Association of Fire Chiefs

EXAM PREP

Fire Fighter I & II

By Dr. Ben A. Hirst,
Performance Training
Systems

JONES AND BARTLETT PUBLISHERS
Sudbury, Massachusetts
BOSTON   TORONTO   LONDON   SINGAPORE

Jones and Bartlett Publishers
World Headquarters
40 Tall Pine Drive
Sudbury, MA 01776
978-443-5000
www.jbpub.com

International Association of Fire Chiefs
4025 Fair Ridge Drive
Fairfax, VA 22033
www.IAFC.org

Performance Training Systems, Inc.
760 U.S. Highway One, Suite 101
North Palm Beach, FL 33408
www.FireTestBanks.com

Jones and Bartlett Publishers Canada
6339 Ormindale Way
Mississauga, Ontario L5V 1J2
Canada

Jones and Bartlett Publishers International
Barb House, Barb Mews
London W6 7PA
United Kingdom

Jones and Bartlett's books and products are available through most bookstores and online booksellers. To contact Jones and Bartlett Publishers directly, call 800-832-0034, fax 978-443-8000, or visit our website www.jbpub.com.

Substantial discounts on bulk quantities of Jones and Bartlett's publications are available to corporations, professional associations, and other qualified organizations. For details and specific discount information, contact the special sales department at Jones and Bartlett via the above contact information or send an email to specialsales@jbpub.com.

Editorial Credits
Author: Dr. Ben A. Hirst

Production Credits
Chief Executive Officer: Clayton E. Jones
Chief Operating Officer: Donald W. Jones, Jr.
President: Robert W. Holland, Jr.
V.P., Sales and Marketing: William J. Kane
V.P., Production and Design: Anne Spencer
V.P., Manufacturing and Inventory Control: Therese Bräuer
Publisher, Public Safety Group: Kimberly Brophy
Associate Managing Editor: Erin Roberts
Production Editor: Karen Ferreira
Director of Marketing: Alisha Weisman
Cover Design: Kristin Ohlin
Interior Design: Anne Spencer
Composition: Northeast Compositors
Printing and Binding: Courier Stoughton

Copyright © 2005 by Jones and Bartlett Publishers and Performance Training Systems, Inc.

Photographs of wet barrel hydrant (pages 112 and 134) and dry barrel hydrant (pages 115 and 138) courtesy of American AVK Company.

ISBN: 0-7637-2847-0

All rights reserved. No part of the material protected by this copyright notice may be reproduced or utilized in any form, electronic or mechanical, including photocopying, recording, or by any information storage and retrieval system, without written permission from the copyright owner.

The procedures in this text are based on the most current recommendations of responsible sources. The publisher and Performance Training Sysyems, Inc. make no guarantees as to, and assume no responsibility for the correctness, sufficiency, or completeness of such information or recommendations. Other or additional safety measures may be required under particular circumstances. This text is intended solely as a guide to the appropriate procedures to be employed when responding to an emergency. It is not intended as a statement of the procedures required in any particular situation, because circumstances can vary widely from one emergency to another. Nor is it intended that this text shall in any way advise firefighting personnel concerning legal authority to perform the activities or procedures discussed. Such local determination should be made only with the aid of legal counsel.

Printed in the United States of America
09 08 07 06 05   10 9 8 7 6 5 4 3 2

CONTENTS

Preface v

Personal Progress Plotter ix

Phase I 1

Examination I-1 3

Examination I-2 26

Examination I-3 49

Feedback Step 80

Phase II 83

Examination II-1 85

Examination II-2 100

Examination II-3 116

Feedback Step 140

Phase III 143

Phase IV 149

Appendix A—Examination I-1 153

Appendix A—Examination I-2 176

Appendix A—Examination I-3 198

Appendix B—Examination II-1 229

Appendix B—Examination II-2 244

Appendix B—Examination II-3 259

PREFACE

The Fire and Emergency Medical Service is facing one of the most challenging periods in its history. Local, state, provincial, national, and international government organizations are under pressure to deliver ever-increasing services. The events of September 11, 2001, continued activities and threats by terrorist organizations worldwide, and the need to maximize available funds are part of the reason most Fire and Emergency Medical Service organizations are examining and reinventing their roles.

The challenge of reinventing the Fire and Emergency Medical Service to provide first response efforts includes increasing professional requirements. Organizations such as the National Fire Protection Association (NFPA), National Professional Qualifications Board (Pro Board), International Fire Service Accreditation Congress (IFSAC), the International Association of Fire Chiefs (IAFC), and the International Association of Fire Fighters (IAFF) are having a dramatic influence on raising the professional qualifications of the first line of defense for emergency response.

Qualification standards have been improved. Accreditation of training and certification are at the highest levels ever in the history of the Fire and Emergency Medical Service. These improvements are reflected in a better prepared first responder, but are not without an effect on those individuals who serve. Fire fighters are being required to expand their roles, acquire new knowledge, develop new and higher level technical skills, as well as participate in requalification and in-service training programs on a regular basis.

The aftermath of September 11, 2001 has had a profound effect on the Fire and Emergency Medical Service. Lessons learned, new technologies, and a national focus on terrorism and weapons of mass destruction are placing much greater demands on fire fighters to keep abreast of change in their specialty operations and improve their technical competence in new technology that was not even available just a few years ago.

Fire fighters cannot afford to be complacent and continue to perform in the same way. Obvious dangers faced by first responders under current heightened security conditions require many adjustments in what is being taught to fire fighters as they operate in an emergency environment. Processes and modes of operation must be carefully examined and must be continuously monitored, changed, and updated.

National leaders are constantly pointing to the first responders as our "first line of defense" against acts of terror and defense of life and property from dangerous weapons that have not been used extensively in our history. Firefighting is highly steeped in tradition. We must question our traditions and our traditional thinking to bring our knowledge, skills, and abilities in line with the demands of today's real world.

Many things have been learned from the September 11, 2001 attacks on America. Some of these lessons learned were the result of our reluctance to change processes and procedures, i.e. our traditions. As great as our tradition is, we in the Fire and Emergency Medical Service must not stop reflecting on the paramount reasons for our existence, which is to protect property, save lives, and perform our tasks with personal safety as the number one concern. These are very important reasons to exist, to improve, and to move from a good Fire and Emergency Service industry to a great Fire and Emergency Service industry.

A word about knowing your business. There are organizations that focus a lot of their training time and effort on the performance side of firefighting. That is essential and is the bottom line for developing skilled fire fighters. The dark side of this approach to training is a lack of emphasis on key knowledge requirements. Often, it is not what we did or did not do as a fire fighter, but what we could have done had we a strong base of knowledge that helps to analyze and detect a need for action outside the routine tasks of firefighting. The Fire and Emergency Medical Service as a whole, and each fire fighter in particular, must focus equally on the knowledge portion of firefighting to help improve the performance side of our tasks. Our fire officers, fire fighters, and support personnel must develop a solid knowledge base so that better judgments, size-ups, and fireground actions can be made. Research in education and training over the years has concluded that lack of knowledge is one of the key reasons why tasks are poorly performed or performed in a manner that did not achieve the expected results.

Never in the history of the Fire and Emergency Medical Service has it become so clear that learning is a career-long, life-long requirement that will be the foundation for the demands that lie ahead. We in the Fire and Emergency Medical Service must adopt this principle to move from a tradition-rich past to become a truly great provider and protector available to everyone we serve. We cannot correct the mistakes of the past, but we can use lessons learned to prevent similar mistakes in the future. Knowledge is power. Efficient and effective people are the solution to moving from a good fire and emergency medical service to a truly great one.

Fire fighters generally do not like to take examinations. For that matter, few people really like them. The primary purposes of the *Exam Prep* Series are to help Fire and Emergency Medical Service personnel develop an improved level of knowledge, eliminate examination taking fear, build self confidence, and develop good study and information mastery skills.

Performance Training Systems, Inc. (PTS) has emerged over the past 16 years to become the leading provider of valid testing materials for certification, promotion, and training for fire and emergency medical personnel. More than 30 examination banks provide the basis for validated examinations. All products are based on the NFPA Professional Qualifications Standards for fire personnel and the Department of Transportation (DOT) Curriculum for emergency medical personnel.

Over the past seven years, PTS has conducted research supporting the development of the Systematic Approach to Examination Preparation® (SAEP). The SAEP has resulted in consistent improvement in scores for persons taking certification, promotion, and training completion examinations. This *Exam Prep* Series is designed to assist fire fighters to improve their knowledge, skills, and abilities while seeking training program completion, certification, and promotion. Using the features of SAEP, coupled with helpful examination-taking tips and hints, will help ensure improved performance from a more knowledgeable and skilled fire fighter.

All examination questions used in SAEP were written by fire and emergency service personnel. Technical content was validated through the use of current technical textbooks and other technical reference materials. Job content was validated by use of technical review committees representative of the Fire Fighter I and II ranks in the fire service, training, and certification organizations. The examination questions represent an approximate 50% sample of the Fire Fighter I and II Test Banks developed and maintained by PTS over the past 16 years. These testing materials are being used by 60 fire service certification agencies worldwide, 94 fire academies, and more than 300 fire department training divisions. Forty-four of the 50 state fire service certification agencies use these testing materials in their programs of certification. For more information on the number of available examination banks and the processes of validation, visit *www.firetestbanks.com*.

Introduction to the Systematic Approach to Examination Preparation

How does SAEP work? SAEP is an organized process of carefully researched phases that permits each person to proceed in examination preparation at an individual's own pace. At certain points self study is required to move from one phase of the program to another. Feedback on progress is the basis of SAEP. It is important to follow the program steps <u>carefully</u> to realize the full benefits of the system.

SAEP allows you to prepare for your next comprehensive training, promotional, or certification examination. Just follow the steps to success. Performance Training Systems, Inc., the leader in producing promotional and certification exams for the fire and emergency medical service industry for over 16 years, has the experience and testing expertise to help you meet your professional goals.

Using the *Exam Prep* manual will enable you to pinpoint areas of weakness in terms of NFPA Standard 1001, and the feedback will provide the reference and page number to help you research the questions that you miss or guess using current technical reference materials. This program is a three-examination set for Fire Fighters as described in NFPA 1001, *Standard for Fire Fighter Professional Qualifications*, 2002 Edition.

Primary benefits of the SAEP include:

- Emphasis on areas of weakness
- Immediate feedback
- Saving time and energy
- Learning technical material through context and association
- Helpful examination preparation practices and hints

Phases of SAEP

SAEP is organized in four distinct Phases for Fire Fighter I and II. The Phases are briefly described as follows.

Phase I

This Phase includes three exams containing items that are selected from each major part of NFPA Standard 1001 for Fire Fighter I Level. An essential part of the SAEP design is to survey your present level of knowledge and build upon it for subsequent examination and self-directed study activities. Therefore, it is suggested that you read the reference materials but <u>do not study</u> or look up any answers while taking the initial examination. Upon completion of the initial examination, you will complete a feedback activity and record examination-items that you missed or that you guessed. We ask you to perform certain tasks during the feedback activity. Once you have completed the initial examination and have researched the answers for any questions you missed, you may proceed to the next examination. The process is repeated through and including the third examination in the Fire Fighter I and II series depending on the level of certification you are seeking.

Phase II

Fire Fighter II examinations are provided for use in this Phase of the SAEP. This Phase includes three exams, each made up of examination items from all Fire Fighter II sections of NFPA 1001, 2002 Edition.

The examinations should be completed as prescribed in the directions supplied with the examination. Complete the feedback report using the same procedures for the Phase I examinations. Pay particular attention to those areas of the references covering material where you score the lowest. At this point, it is important to read the materials containing the correct response in context once again. This technique will help you master the material, relate it to other important information, and retain knowledge.

Phase III

Phase III contains important information about examination-item construction. It provides insight regarding the examination-item developers, how they apply their technology, and hints and tips to help you score higher on any examination. Make sure you read this Phase carefully. It is a good practice to read it twice, and study the information a day or two prior to your scheduled examination.

Phase IV

Phase IV information addresses the mental and physical aspects of *Exam Prep*. By all means, do not skip this part of your preparation. Points can be lost if you are not ready, physically and mentally, for the examination. If you have participated in sporting or other competitive events, you know the importance of this level of preparation. There is no substitute for readiness. Just being able to answer the questions will not move you to a level of excellence and to the top of the examination list for training, promotion, or certification. Quality preparation is much more than just answering examination items.

Supplemental Practice Examination Program

The supplemental practice examination program differs from the SAEP program. It is provided over the Internet 24 hours a day, 7 days a week. This supplemental practice examination allows you to make final preparations immediately before your examination date. You will get an immediate feedback report that includes the questions missed and the references/page numbers for those missed questions. The practice examination will help you concentrate on areas of greatest weakness and will save you time and energy immediately before the examination date. If you choose this method, **do not** "cram" for the examination. The upcoming helpful hints for examination preparation will explain the reasons for avoiding a "cramming exercise." A supplemental practice examination is available as a part of the cost of this *Exam Prep* manual by using the enclosed registration form. Do not forget to fax a copy of your Personal Progress Plotter along with your Registration Form. The data supplied on your Personal Progress Plotter will be kept confidential and will be used by PTS to make future improvements in the *Exam Prep* Series. You may take a short practice examination to get the procedure clear in your mind by going to *www.webtesting.cc*.

Good luck in your efforts to improve your knowledge and skills. Our primary goal is to improve the Fire and Emergency Medical Service one person at a time. We want your feedback and impression of the System to help us implement improvements in future editions of the *Exam Prep* Series of books. Address your comments and suggestions to *www.firetestbanks.com*.

Rule 1
Examination preparation is not easy. Preparation is 95% perspiration and 5% inspiration.

Rule 2
Follow the steps very carefully. Do not try to reinvent or shortcut the system. It really works just as it was designed to!

Personal Progress Plotter

Fire Fighter I Exam Prep

Name: _____

Date Started: _____

Date Completed: _____

Fire Fighter II Exam Prep

Name: _____

Date Started: _____

Date Completed: _____

Fire Fighter I	Number Guessed	Number Missed	Examination Score
Examination I-1			
Examination I-2			
Examination I-3			

Fire Fighter II	Number Guessed	Number Missed	Examination Score
Examination II-1			
Examination II-2			
Examination II-3			

Formula to compute Examination Score = (Number guessed + Number missed) × Point Value per examination item subtracted from 100.

Note: 200-Item Examination = .5 point per examination item
150-Item Examination = .67 points per examination item

Example: Examination I-1, 5 examination items were guessed, 8 were missed for a total of 13 on a 100-item examination. The Examination Score would be 100 − (13 × 1.0 points) = 87

Example: Examination II-1, 5 examination items were guessed, 8 were missed for a total of 13 on a 150-item examination. The Examination Score would be 150 − (13 × .67 points) = 91.29

Note: In order to receive your free online practice examination, you must fax a copy of your completed personal progress plotter along with your registration form.

PHASE I

Fire Fighter I

Examination I-1, Beginning NFPA 1001

Taking the 150-Item Examination – Examination I-1. **Do not** study prior to taking the examination. The examination is designed to identify your weakest areas in terms of NFPA 1001, 2002 Edition. There will be steps in the SAEP that require self-study of specific reference materials. Remove Examination I-1 from the book. Mark all answers in ink. The reason for this is to make sure no changes are made. **Do not** mark through answers or change answers in any way once you have selected the answer.

 Step 1—Take Examination I-1. When you have completed Examination I-1 go to Appendix A and compare your answers with the correct answers. Notice that each answer has reference materials with page numbers. If you missed the correct answer to the examination item, you have a source for conducting your correct answer research.

 Step 2—Score Examination I-1. How many examination items did you miss? Write the number of missed examination items in the blank **in ink**. _____ Enter the number of examination items you guessed in this blank. _____ Go to your personal progress plotter following the Introduction and enter these numbers in the designated locations.

 Step 3—Now the learning begins! Carefully research the page cited in the reference material for the correct answer. For instance, use Jones and Bartlett *Fundamentals of Fire Fighter Skills*, First Edition, go to the page number provided, and find the answer.

Rule 3

Mark with an "X" any examination items for which you guessed the answer. For maximum return on effort, you should also research any answer that you guessed, even if you guessed correctly. Find the correct answer, highlight it, and then read the entire paragraph that contains the answer. Be honest and mark all questions you guessed. Some examinations have a correction for guessing built into the scoring process. The correction for guessing can reduce your final examination score. If you are guessing, you are not mastering the material.

Helpful Hint

Most of the time your first impression is the best. More than 41% of answers changed during our SAEP field test were changed from a right answer to a wrong answer. Another 33% changed their answer from wrong answer to wrong answer. Only 26% of changed answers were from a wrong answer to a right answer. In fact, a number of changed answers resulted in the participant not making a perfect score of 100%! Think twice before you change your answer. The odds are not in your favor.

Helpful Hint

Researching correct answers is one of the most important activities in SAEP. Locate the correct answer for any item answered incorrectly. Highlight the correct answer. Then read the entire paragraph containing the answer. This will put the answer in context for you and provide important learning by association.

Helpful Hint

Proceed through all missed examination items using the same technique. Reading the entire paragraph improves retention of the information and helps you develop an association with the material and learn the correct answers. This step may sound simple. A major finding during the development and field testing of SAEP was that you learn from your mistakes.

Examination I-1

Directions

Remove Examination I-1 from the manual. First, take a careful look at the examination. There should be 150 examination items. Notice that a blank line precedes each examination item number. This line is provided for you to enter the answer to the examination item. Write the answer **in ink**. Remember the rule about changing the answer. Our research has shown that changed answers are often incorrect, and more often than not the answer that is chosen first is correct.

If you guess the answer to a question, place an "X" or a checkmark by your answer. This step is vitally important as you gain and master knowledge. We will explain how we treat the "guessed" items later in SAEP.

Take the examination. Once you complete it, go to Appendix A and score your examination. Once the examination is scored, carefully follow the directions for feedback on the missed and guessed examination items.

_____ 1. Life safety, incident stabilization, and _____ are the three most important organizational duties for fire departments to pursue.
   A. building inspections
   B. public information
   C. property conservation
   D. resource management

_____ 2. The plan or written document for tactical operations is known as a department's:
   A. S.O.P./S.O.G.
   B. organizational chart.
   C. prefire plan.
   D. mission statement.

_____ 3. Fire department standard operating procedures should be established in the **most commonly** accepted order of fireground priorities, which is:
   A. life safety, property conservation, fire control.
   B. property conservation, life safety, fire control.
   C. fire control, life safety, property conservation.
   D. life safety, incident stabilization, and property conservation.

_____ 4. The **most common** injuries related to improper lifting techniques are:
   A. back strains.
   B. bruises.
   C. sprains.
   D. fractures.

_____ 5. In the Incident Management System (IMS), the functional area responsible for all costs and financial aspects of an incident, especially at a large-scale, long-term incident, is:
   A. Finance.
   B. Operations.
   C. Logistics.
   D. Command.

_____ 6. Depending on a department's mission statement, a fire fighter may also be required to perform as a(n):
   A. Hazardous Materials Technician.
   B. Rescue Specialist.
   C. Emergency Medical Technician or Paramedic.
   D. All of the above

_____ 7. In the Incident Management System (IMS), the functional area responsible for providing facilities, services, and materials necessary to support an incident is:
   A. Planning.
   B. Operations.
   C. Logistics.
   D. Command.

_____ 8. One of the **primary** functions of the truck/ladder company is:
   A. performing forcible entry to fire building.
   B. directing traffic at fire scene.
   C. laying supply lines for engine companies.
   D. fire attack.

_____ 9. The **primary** use of the _____ is to attach a rope to a cylindrical object.
   A. clove hitch
   B. bowline
   C. becket/sheet bend
   D. rescue knot

_____ 10. The _____ is tied as shown below:
   A. bowline
   B. becket/sheet bend
   C. clove hitch
   D. figure-eight

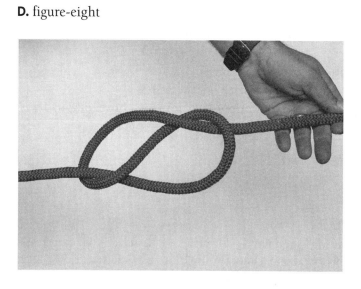

_____ 11. The part of the rope that is used for work such as hoisting or pulling is called the:
   A. working end.
   B. round turn.
   C. running end.
   D. standing part.

_____ 12. Fire service rope falls into two use classifications:
   A. life safety and utility.
   B. braided and kernmantle.
   C. dynamic and static.
   D. natural and synthetic.

_____ 13. The knot used to tie two ropes of unequal diameter together is the:
   A. clove hitch.
   B. square knot.
   C. becket/sheet bend.
   D. half hitch.

_____ 14. The _____ is tied as shown below.
   A. bowline
   B. becket/sheet bend
   C. clove hitch
   D. figure-eight

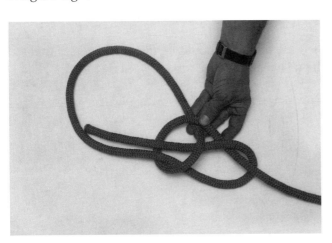

_____ 15. The clove hitch is commonly used in the fire service for:
   A. stringing lines together.
   B. forming the rescue knot.
   C. hoisting fire equipment.
   D. splicing rope together.

_____ 16. A knot forming a loop that will not slip under strain and can be untied easily is a:
   A. timber hitch.
   B. half hitch.
   C. figure-eight on a bight.
   D. half sheep shank with a safety.

_____ 17. The _____ is tied as shown below.
   A. bowline
   B. becket/sheet bend
   C. square knot
   D. follow through

_____ 18. When hoisting a ladder, the rope should be threaded and secured through the ladder _____ the distance from the top.
   A. 1/6
   B. 1/5
   C. 1/3
   D. 1/2

_____ 19. The _____ is tied as shown below.
   A. bowline
   B. half-hitch
   C. clove hitch
   D. figure-eight

20. Which of the following has the highest priority for a radio transmission?
   A. Notification from dispatch of road closures
   B. Emergency traffic from a unit working at a fire or rescue
   C. Vital signs of a patient being transported by fire department ambulance
   D. Transmission of local forest fire danger notice

21. All fire department radio operations must follow rules of the:
   A. NFPA.
   B. DOT.
   C. FCC.
   D. CFR.

22. Computer-aided dispatch (CAD) can be defined as a(n):
   A. computer-based automated system that assists the telecommunicator in assessing dispatch information and recommends responses.
   B. organized collection of similar facts.
   C. system typically used by operations chief officers in the fire service.
   D. emergency alerting devices primarily used by volunteer department personnel to receive reports of emergency incidents.

23. When receiving reports of emergencies by telephone, the individual should always speak:
   A. rapidly with low volume.
   B. softly with some hesitation.
   C. clearly, slowly, and with good volume.
   D. clearly, rapidly, and with good volume.

24. When speaking to the public, the individual receiving the call should:
   A. use as many technical terms as possible.
   B. use plain everyday language.
   C. disregard callers who scream hysterically into the phone.
   D. argue with known abusers of the 911 system.

25. Before transmitting any information over the fire department radio, a fire fighter should:
   A. press the key two or three times to signal the intent to transmit.
   B. key the microphone and then clear his/her throat to be sure his/her voice will be clear.
   C. turn up the volume control if his/her voice is naturally soft.
   D. listen to be sure the channel is not being used.

26. The coupling on the high pressure hose that is attached to an SCBA cylinder should be:
   A. of the reverse thread type.
   B. hand tight.
   C. secured with an adjustable wrench.
   D. treated with teflon tape.

_____ 27. When using positive pressure SCBA, a poor seal between the facepiece and the fire fighter's face is:
   A. not dangerous, because the positive pressure will keep toxic gases out of the facepiece.
   B. dangerous, because it is still possible for toxic gases to enter the facepiece.
   C. not possible, because the positive pressure will seal the facepiece to the face.
   D. the main cause of deaths on the fireground.

_____ 28. All of the following are good practices when wearing self-contained breathing apparatus, <u>except</u>:
   A. working in pairs as a minimum.
   B. checking facepiece seal prior to connecting the breathing tube.
   C. turning the bypass valve on when entering the structure.
   D. leaving a dangerous area immediately when alarm sounds.

_____ 29. The bypass valve on a self-contained breathing apparatus is used:
   A. during hazardous materials incidents.
   B. in emergency situations involving a malfunctioning regulator.
   C. to clear the mask of unwanted condensation.
   D. to cool the face piece when high heat is encountered.

_____ 30. Which of the following is considered to be a hazardous atmosphere encountered during fires?
   A. Oxygen deficiency
   B. Elevated temperatures
   C. Smoke
   D. All of the above

_____ 31. When tightening the straps on a SCBA facepiece, the _____ straps should always be tightened first.
   A. lower
   B. temple
   C. upper
   D. harness

_____ 32. As the oxygen supply in any given area falls below _____ percent, unconsciousness can occur.
   A. 21
   B. 17
   C. 12
   D. 9

_____ 33. The <u>first</u> noticeable signs of oxygen deficiency are:
   A. profuse sweating and ringing in the ears.
   B. dizziness, impaired vision, and giddiness.
   C. increased respiratory rate and impaired muscular coordination.
   D. headache and rapid fatigue.

____ 34. Four hazardous atmospheres that fire fighters are likely to encounter at a fire are:
   A. super-heated air, toxic gases, oxygen deficiency, and smoke.
   B. toxic gases, hyperventilation, oxygen deficiency, heat.
   C. heat, light, smoke, and chemical chain reaction.
   D. carbon monoxide, ammonia, water, and hydrogen sulfide.

____ 35. The open-circuit, positive-pressure breathing apparatus operates by:
   A. maintaining a pressure inside the mask that is slightly higher than atmospheric pressure.
   B. generating oxygen for rebreathing by a chemical reaction within the breathing apparatus.
   C. providing pure oxygen to the user.
   D. blending pure oxygen with recirculating exhaled air.

____ 36. Rescue from sewers, grain silos, and similarly confined spaces requires the use of self-contained breathing apparatus due to the danger of:
   A. toxic gases.
   B. oxygen deficiency.
   C. ambient temperature.
   D. Both A and B are correct.

____ 37. Trapped fire fighters awaiting rescue will use less air if they:
   A. partially close the cylinder valve.
   B. open the bypass valve.
   C. struggle to get free.
   D. control their breathing.

____ 38. The **primary** function of the bypass valve on SCBA is:
   A. to enable the wearer to breathe more oxygen.
   B. for use if the regulator fails.
   C. to help control excessive heat in the facepiece.
   D. to facilitate removal of condensation from the lens of the facepiece.

____ 39. The proper position of the bypass valve on positive-pressure SCBA under normal conditions is:
   A. fully open.
   B. cracked open.
   C. fully closed.
   D. open three full turns.

____ 40. All SCBA regulators should be provided with a _____ valve to be opened in the event of regulator failure.
   A. mainline
   B. reducing
   C. bypass
   D. safety

_____ **41.** A fire fighter is susceptible to poisoning or irritation from carbon monoxide through:
A. ingestion.
B. absorption.
C. inhalation.
D. injection.

_____ **42.** Which type of breathing apparatus recycles the user's exhaled breath after removing carbon dioxide and adding supplemental oxygen?
A. Open-circuit
B. Closed-circuit
C. SAR
D. Positive pressure

_____ **43.** Riding on a tailboard of an apparatus is:
A. acceptable if an enclosed cab is not available.
B. acceptable if the fire fighter is wearing a safety belt.
C. a major safety violation.
D. necessary for transporting fire fighters to the scene.

_____ **44.** When an enclosed cab **is not** available and the fire fighter must ride in a jump seat, the fire fighter should use:
A. a safety bar.
B. eye protection.
C. ear protection.
D. All of the above.

_____ **45.** Failure to wear your PPE can lead to:
A. negligence.
B. injury.
C. fines.
D. increased protection.

_____ **46.** NFPA requires that, as a minimum, PPE clothing should be cleaned every _____ months.
A. 12
B. 6
C. 9
D. 14

_____ **47.** Which of the following **is not** one of the components of the layered protection system in structural PPE?
A. Vapor/moisture barrier
B. Waterproof outer layer
C. Thermal barrier
D. Outer shell

_____ 48. To improve dexterity with gloves, be sure to have a good fit and:
   A. purchase the most expensive gloves available.
   B. use the gloves only when required.
   C. practice with the gloves.
   D. purchase gloves that are advertised in trade journals.

_____ 49. A simple problem in using the non-integrated PASS devices is that it has contributed to numerous fire fighter fatalities due to:
   A. forgetting to turn on the unit.
   B. neglecting to attach antennae to homing unit.
   C. the fact that the device is too technical for most fire fighters.
   D. the fact that the device is over-pressurized.

_____ 50. All of the following concerning downed power lines are correct **except**:
   A. downed power lines should be considered energized until the power company confirms it is dead.
   B. secure the area around the power line and keep the public at a safe distance.
   C. never drive fire apparatus over a downed power line.
   D. it is acceptable to drive apparatus over a downed line because the rubber in the tires acts as an insulator.

_____ 51. A flat-head axe is more suitable for _____, while a pick-head axe is more adaptable to a variety of firefighting functions.
   A. striking
   B. prying
   C. heavy work
   D. chopping

_____ 52. Tempered plate glass should be broken only as a last resort. It is recommended that it be shattered by striking:
   A. in the center with a large, blunt object.
   B. at the lowest corner with the pick end of a fire axe.
   C. with a flat-headed axe with hands above the head.
   D. with a flat-headed axe at the highest corner.

_____ 53. In what type of occupancy will fire fighters usually find sliding doors?
   A. Barns or warehouses
   B. Commercial
   C. Residential
   D. Institutional

_____ 54. Firefighters can expect to find ledge doors in:
   A. barns and warehouses.
   B. single-family residential buildings.
   C. churches and temples.
   D. commercial occupancies.

_____ 55. A battering ram is intended to be used by two or four fire fighters. Which of the following describes its recommended use?
   A. It is held horizontally by all involved fire fighters, who run toward the object to be battered.
   B. If four fighters are involved, one guides the tip of the ram while two swing it in the direction of the fourth person.
   C. It is held horizontally by pairs of operators who swing it repeatedly against the object.
   D. It is held against the object to be battered by two fire fighters, while the third person pounds on the end with a sledge hammer.

_____ 56. The **primary** rule of forcible entry is:
   A. "Look before you leap."
   B. "Try before you pry."
   C. "Always stand to leeward."
   D. "Sharp edge pointed down and ahead."

_____ 57. The K-tool is useful in:
   A. opening roofs.
   B. opening walls.
   C. pulling lock cylinders.
   D. breaking window glass.

_____ 58. Because of the high risks associated with search and rescue:
   A. all floors must be searched simultaneously.
   B. it is done by teams of fire fighters.
   C. one fire fighter should accomplish the task.
   D. it is not recommended in large structures.

_____ 59. Safety requires that fire fighters using self-contained breathing apparatus must work:
   A. alone.
   B. in contained areas.
   C. in pairs.
   D. with a lifeline.

_____ 60. Before fire fighters enter a burning building to perform rescue work, they must first consider:
   A. manpower on the scene.
   B. weather conditions.
   C. damaging evidence of forced entry.
   D. their own safety.

_____ 61. One purpose of a Personnel Accountability System is to:
   A. keep track of fire fighters' salaries.
   B. identify trapped or injured fire fighters.
   C. create a job complaint forum.
   D. help the Fire Chief control sick leave abuse.

_____ **62.** Using a ladder at a reduced angle (less than 75°) drastically reduces the _____ of the ladder and _____ the possibility of slippage.
  A. weight, reduces
  B. load-carrying capacity, increases
  C. curvature, increases
  D. width, reduces

_____ **63.** The parts of an extension ladder that prevent the fly section from being extended too far are called:
  A. guides.
  B. locks.
  C. anchors.
  D. stops.

_____ **64.** The dogs of an extension ladder serve to:
  A. prevent the fly section from being over-extended.
  B. hold the fly section in place after it has been extended.
  C. raise the fly section.
  D. guide the fly section while it is being raised.

_____ **65.** A bangor ladder has attachments for added leverage that are called:
  A. staypoles.
  B. truss poles.
  C. guide poles.
  D. rails.

_____ **66.** The proper name for one of the two principal structural sides of a ladder is a:
  A. truss block.
  B. beam.
  C. guide.
  D. rung.

_____ **67.** When referring to ladders, the term tie rod is used to describe a metal rod running from:
  A. one beam to the other.
  B. a beam to a rung.
  C. rung to truss block.
  D. the pulley(s) to the ladder frame.

_____ **68.** When approaching a fire in the passenger compartment of a vehicle, the **best** stream pattern to use is a _____ stream.
  A. straight
  B. narrow fog
  C. wide fog
  D. solid

_____ 69. <u>Directions</u>: Read the statements below, then select your answer from choices A-D.

The basic procedure for extinguishing a vehicle fire is:

1. First, attack the fire in the vehicle.
2. Next, extinguish any ground fire around the vehicle.
3. Approach vehicle from side, being cautious around wheel/tire areas.

    A. Statement 1 is true; statements 2 and 3 are false.
    B. All three statements are true.
    C. Statement 1 is false; statements 2 and 3 are true.
    D. Statements 1 and 2 are false, statement 3 is true.

_____ 70. Which of the following is a prying tool that could be used on a vehicle fire?
    A. Halligan
    B. Ram bar
    C. Plaster hook
    D. Flathead axe

_____ 71. The safest way to attack a fully involved vehicle fire is from the _____ side when possible.
    A. upwind and uphill
    B. downwind and uphill
    C. down hill and leeward
    D. leeward and level ground

_____ 72. What is the method of fire suppression **most** widely used on Class A combustibles?
    A. Cooling
    B. Removal of fuel
    C. Smothering
    D. Inhibition of chain reaction

_____ 73. Water is the <u>primary</u> fire extinguishing agent because of its:
    A. temperature coefficient and surface tension.
    B. purity and low toxicity.
    C. ability to absorb heat.
    D. flexibility and tendency to evaporate slowly.

_____ 74. A typical siamese appliance will have _____ connections:
    A. 3 male
    B. 2 male and 1 female
    C. 2 female and 1 male
    D. 3 female

_____ 75. In the fire service, the basic definition of the word rescue is:
    A. removing a victim from a hazardous situation to safety.
    B. stabilizing a victim before transporting.
    C. performing CPR on a victim.
    D. All of the above.

___ 76. During search operations, the fire fighter should occasionally pause and:
   A. rest to conserve air.
   B. regroup with all those searching.
   C. listen for calls or signals for help.
   D. leave a trail to aid escape.

___ 77. Which of the following are characteristics of a **primary** search?
   A. Rapid and systematic
   B. Slow and deliberate
   C. Accomplished only with a charged hose line
   D. Begun only after top-side ventilation is provided

___ 78. Which of the following can indicate the possibility of a building collapse?
   A. Loose mortar joints
   B. Leaning of exterior walls
   C. Creaks and cracking sounds
   D. All of the above

___ 79. When exposed to products of combustion, the _____ are more vulnerable to injury than any other body area.
   A. lungs and respiratory tract
   B. heart and respiratory tract
   C. lungs and eyes
   D. brain and spinal cord

___ 80. What type of attack is typically used by fire fighters in structural fires?
   A. Master streams
   B. Direct
   C. Combination
   D. Indirect

___ 81. A hose roller is a:
   A. rope or strap used to hoist hose.
   B. device by which a hose can be rolled.
   C. device that is fastened to the hose coupling in preparation for fastening.
   D. curved metal frame that fits over the edge of a roof or windowsill.

___ 82. Solid-stream handlines are designed to be operated at a nozzle pressure of _____ psi.
   A. 50
   B. 75
   C. 90
   D. 100

___ 83. When fire fighters are unable to enter the structure or fire area due to intense fire conditions, a(n) _____ attack can be made.
   A. direct
   B. blitz
   C. combination
   D. indirect

____ 84. The **most efficient** use of water on localized interior fires is made by a(n) _____ attack on the base of the fire.
   A. transitional
   B. direct
   C. indirect
   D. combination

____ 85. When a hose is being rolled into twin rolls secured together by a portion of the hose itself, it is called a _____ roll.
   A. donut
   B. hi-rise
   C. self-locking
   D. straight

____ 86. Prior to entering a fire area with a charged hoseline, the nozzle operator should first:
   A. bleed air from the line.
   B. wait for a building layout.
   C. wait for the power to be shut off.
   D. wait for direction from the pump operator.

____ 87. In a multi-story building with a standpipe system, a fire fighter should make the connection _____ the fire floor.
   A. on the floor below
   B. on the floor above
   C. on
   D. two levels above

____ 88. In reference to a fire stream, friction loss is defined as:
   A. the loss of the stream velocity after it exits the nozzle.
   B. that part of the pressure that is used to overcome friction in the hose.
   C. the amount of pressure needed to overcome friction caused by water turbulence.
   D. the friction created by dragging the hose along the ground.

____ 89. When water flowing through a fire hose or pipe is suddenly stopped, the resulting surge is referred to as:
   A. static energy absorption.
   B. a water hammer.
   C. flow pressure.
   D. residual pressure.

____ 90. To properly operate a fire hose nozzle valve, it should be:
   A. opened slowly and closed quickly.
   B. opened quickly and closed slowly.
   C. opened and closed slowly.
   D. opened and closed quickly.

91. _____ nozzle will discharge a wide range of flows with an effective fire stream, depending on the pressure being supplied to the nozzle.
  A. Set gallonage
  B. Solid
  C. Automatic
  D. Proportioner

92. A stream discharging <u>more</u> than 300 gpm is know as a(n) _____ stream.
  A. booster
  B. solid
  C. elevated
  D. master

93. A fire stream used in an indirect attack in a structure fire should be shut down before:
  A. all spot fires are extinguished.
  B. a backup line becomes available.
  C. ventilation occurs.
  D. thermal layering is disturbed.

94. Proper ventilation reduces danger of asphyxiation, enhances visibility, and removes:
  A. overhaul concerns.
  B. excess moisture.
  C. salvage concerns.
  D. heat.

95. The side of a building that the wind is striking is called the _____ side. The opposite side is called the _____ side.
  A. windward, leeward
  B. leeward, windward
  C. downwind, upwind
  D. upwind, downwind

96. The removal of smoke from windowless basement areas often requires the use of:
  A. natural openings.
  B. mechanical ventilation.
  C. vertical ventilation.
  D. horizontal ventilation.

97. Proper ventilation results in an orderly movement of _____ through and out of the structure.
  A. water fog
  B. hose line crews
  C. rescue personnel
  D. heated fire gases

_____ 98. The **primary** function of smoke ejectors or exhaust fans is:
   A. localizing the fire.
   B. removing heat and smoke.
   C. providing fresh air for attack crews.
   D. removing lighter-than-air gases.

_____ 99. Forced/mechanical ventilation is accomplished by blowers, fans, or:
   A. removal of windows.
   B. vertical openings.
   C. fog streams.
   D. natural wind currents.

_____ 100. Convection is:
   A. transfer of heat through space by infrared rays.
   B. transfer of heat through a solid medium.
   C. not considered a method of heat transfer.
   D. transfer of heat through liquids or gases by circulating currents.

_____ 101. Factor(s) that will influence the air currents in a structure fire is/are:
   A. a vertical vent opening.
   B. outside wind direction.
   C. the direction the hose attack team is using in relation to the fire.
   D. All of the above

_____ 102. Combustion is the result of a _____ reaction.
   A. mechanical
   B. chemical
   C. dielectrical
   D. replenishment

_____ 103. When exposed to intense heat, a lightweight metal truss can:
   A. maintain its structural integrity.
   B. often contain a fire to a specific area.
   C. be expected to fail in 5 to 10 minutes.
   D. be expected to support firefighting operations for at least 20 minutes.

_____ 104. Directing fire streams downward through roof openings can cause the heat and smoke to be:
   A. used as an advantage in fire extinguishment.
   B. cooled off and to mushroom throughout the fire building.
   C. discharged to the leeward openings.
   D. forced back into the building, possibly injuring occupants and firefighting personnel.

_____ 105. To be **most effective**, trench ventilation operations must be completed before:
   A. fire attack.
   B. the fire reaches the smoldering stage.
   C. the fire reaches the trench.
   D. extinguishment begins.

____106. When opening roofs, which of the following **is not** a recommended safety practice?
   A. Utilizing natural openings
   B. Cutting large holes rather than small ones
   C. Insuring that main structural supports are not cut
   D. Standing on the leeward side of the hole while working

____107. A power saw should be started on the ground to ensure operation. Before hoisting or carrying to the roof, it should be:
   A. refueled and the blade tightened.
   B. shut off.
   C. tested on available material.
   D. cooled for safety and handling.

____108. When opening a roof, stand _____ the cut.
   A. above
   B. below
   C. on the windward side of
   D. on the leeward side of

____109. The phenomenon by which heat, smoke, and fire gases will travel upward to the highest point and become trapped, accumulate, bank down, and spread out laterally is known as:
   A. backdraft.
   B. mushrooming.
   C. flashover.
   D. buoyancy.

____110. During salvage and overhaul operations, it is essential for fire fighters to:
   A. remove their coats once the fire has been extinguished.
   B. work without coat, gloves, and helmet.
   C. remove protective breathing apparatus.
   D. wear complete protective equipment, including SCBA.

____111. Searching for hidden fires is a **primary** function of:
   A. size up.
   B. salvage.
   C. fire investigation.
   D. overhaul.

____112. Cracked plaster, peeling paint, and discoloration of materials may be signs of:
   A. an intense fire.
   B. fire origin.
   C. possible arson fire.
   D. hidden fire in a wall.

____113. A tool often used to open a ceiling is a:
   A. pick-hand axe.
   B. pike pole.
   C. kelly tool.
   D. clemens hook.

_____ 114. One way to remove water coming through the ceiling from upper floors is by the use of:
   A. sponges.
   B. chutes.
   C. carryalls.
   D. floor runners.

_____ 115. Which of the following is placed on the floor to hold small amounts of water?
   A. Floor runner
   B. Carryall
   C. Catchall
   D. Water chute

_____ 116. A commonly used method for two fire fighters to deploy a large salvage cover is a(n):
   A. combination throw/toss.
   B. balloon throw/toss.
   C. accordion toss.
   D. horseshoe throw/toss.

_____ 117. Fire protection professionals agree that salvage work is:
   A. not practical considering present day staffing requirements.
   B. not a fire department responsibility.
   C. an effective means of promoting positive public relations.
   D. best accomplished by private companies who specialize in this kind of work.

_____ 118. To form the corners of the basin when constructing a catchall, the fire fighter should lay ends of the side rolls over at a _____ angle.
   A. 30°
   B. 60°
   C. 90°
   D. 180°

_____ 119. Methods and operating procedures that reduce fire, water, and smoke damage during and after fires are known as:
   A. overhaul.
   B. size up.
   C. salvage.
   D. a coordinated fire attack.

_____ 120. In salvage operations, floor runners:
   A. are fire fighters who carry debris from the building.
   B. are water chutes constructed of rolled-up salvage covers placed to catch and drain excess water.
   C. is a term used to describe the type of fire which progresses vertically.
   D. are constructed of a lightweight, durable material placed over the floor to protect it from damage.

___ **121.** After connecting the supply line to the hydrant, the fire fighter should _____ the hydrant.
  A. partially open
  B. fully open
  C. barely crack open
  D. never open

___ **122.** When performing a forward lay, the pumper is driven from the:
  A. fire scene to water source.
  B. water source to fire scene.
  C. water source to fire scene to water source.
  D. discharge of one pumper to intake of another.

___ **123.** The term reverse lay describes an apparatus that lays out a supply line:
  A. while the apparatus is moving in reverse.
  B. from the fire to the water source.
  C. from the water source to the fire.
  D. with the male coupling ending up at the fire scene.

___ **124.** Soft sleeve intake hose is used for:
  A. transferring water from a hydrant to an apparatus.
  B. primarily drafting water from a static source.
  C. siphoning water from one portable tank to another.
  D. transferring water from pump to tank.

___ **125.** An advantage of a forward lay is:
  A. that the pumper is located at the hydrant to boost water pressure.
  B. that the supply line is dropped off at the fire location.
  C. the ability to utilize poor or static water sources.
  D. that the pumper is located at the fire with access to additional hoselines.

___ **126.** The space provided for hose on fire apparatus is generally referred to as the:
  A. hose box.
  B. hose load.
  C. hose bed.
  D. hose lay.

___ **127.** AFFF extinguishing agents are applicable to _____ fires.
  A. Class C
  B. Class D
  C. both Class A and B
  D. Class A, B, and C

___ **128.** All portable extinguishers are classified according to their:
  A. size.
  B. freeze potential.
  C. intended use.
  D. conductivity.

___129. A dry chemical extinguisher rated 60-B is capable of extinguishing a _____ flammable liquid pan fire.
A. 40 ft²
B. 60 ft²
C. 120 ft²
D. 150 ft²

___130. CO_2 and dry chemical extinguishers will extinguish both Class B and C fires. What advantage does CO_2 have over a dry chemical extinguisher?
A. CO_2 is not a hazard in an enclosed area.
B. CO_2 does not leave a residue or corrode electrical contacts.
C. CO_2 will prevent reignition longer than dry chemical extinguisher.
D. CO_2 is effective at a greater distance.

___131. A blue circle with a letter designation in the center would indicate an extinguisher is rated for use on _____ fires.
A. Class A
B. Class B
C. Class C
D. Class D

___132. A green triangle containing a letter would indicate an extinguisher to be used on _____ fires.
A. Class A
B. Class B
C. Class C
D. Class D

___133. The proper type of extinguisher for a fire involving magnesium, titanium, or sodium is:
A. dry chemical.
B. CO_2.
C. dry powder.
D. water.

___134. All power outlets used on a scene should:
A. be easily off loaded from the truck.
B. have at least 4 outlets.
C. have ground-fault circuit interrupters.
D. All of the above.

___135. When faced with an electrical emergency the fire fighter shall try to:
A. obtain a pair of lineman gloves.
B. have equipment de-energized.
C. use a dry rope to pull victim from contact with an energized conductor.
D. wear rubber boots when approaching the emergency scene.

____136. Firefighters should treat all downed wires as:
   A. energized.
   B. safe if in contact with the ground.
   C. only dangerous if nearby homes have power.
   D. safe if not arcing.

____137. Which one of the following statements regarding wildland firefighting is **incorrect**?
   A. From a flat to a 30° slope, the fire will double its rate of spread.
   B. A ridge fire does not tend to draw fire to itself.
   C. A fire burns downhill faster than it does uphill.
   D. A fire burns uphill faster than it does downhill.

____138. A suppression action taken by a fire fighter around the perimeter of a wildland fire is called:
   A. black line.
   B. direct attack.
   C. cold fire edging
   D. water curtain.

____139. The three **most important** factors that affect wildland firefighting are:
   A. fuel, equipment, and location.
   B. topography, resources, and time of day.
   C. fuel, weather, and topography.
   D. staffing, resources, and apparatus.

____140. A facility in which there is a great potential likelihood of life or property loss from a fire is called a _____ hazard.
   A. special
   B. assembly
   C. target
   D. industrial

____141. Purposes for fire company surveys include all of the following **except** to:
   A. detect and eliminate hazards.
   B. collect information for prefire planning.
   C. provide a show of force to the public and building owner.
   D. provide valuable life safety information services to property owners.

____142. An effective fire company inspection plan will reap which of the following benefits?
   A. Reduction of fire hazards
   B. Positive public contact
   C. Building familiarization
   D. All of the above.

___143. The goal of fire prevention activities is to accomplish all of the following **except**:
  A. reducing the dangers to fire fighters.
  B. receiving appropriate media coverage of emergencies.
  C. reducing the risk of public safety.
  D. reducing the amount of property damage.

___144. The simplest and **most effective** method of achieving the fire service goal of the preservation of life and property is:
  A. prevention.
  B. improved technology.
  C. more fire fighters.
  D. more fire stations.

___145. One topic a fire fighter may be asked to present during a fire safety presentation to an external group is:
  A. fire stream applications.
  B. home safety practices.
  C. fire extinguisher maintenance.
  D. rescue practices.

___146. When drying synthetic rope, keep it:
  A. in a warm, dry, sunlit place.
  B. in a cool, moist, sunlit place.
  C. out of sunlight.
  D. in a damp, dark place.

___147. Recharging air cylinders can be done from a bank of three or more large air cylinders. This type of system is called a(n) _____ system.
  A. combination
  B. cascade
  C. multiple
  D. in line

___148. A fire department's comprehensive SCBA program should include:
  A. inspecting, disinfecting, maintaining, and storing.
  B. visual inspection of the harness and frame only.
  C. an annual maintenance as a minimum.
  D. using, recording, cleaning, and examining only.

___149. When hose has been exposed to small amounts of oil, it should be washed with:
  A. clear water.
  B. cold water and scrub brush.
  C. mild soap or detergent.
  D. solvent solution.

_____**150.** If dust and dirt on a woven-jacket fire hose cannot be removed by brushing, it should be washed with:
   A. mild soap or detergent.
   B. clear water.
   C. hot, soapy water.
   D. lye soap and water.

Now that you have finished the feedback step for Examination I-1, it is time to repeat the process by taking another comprehensive examination of the NFPA 1001 Standard.

Did you score higher than 80% on Examination I-1? Circle Yes or No <u>in ink</u>.
(We will return to your Yes or No answer to this question later in SAEP)

Examination I-2, Adding Difficulty and Depth

During Examination I-2, progress will be made in developing depth of knowledge and skills.

Step 1—Take Examination I - 2. When you have completed Examination I-2, go to Appendix A and compare your answers with the correct answers.

Step 2—Score Examination I-2. How many examination items did you miss? Write the number of missed examination items in the blank **in ink**. _____ Enter the number of examination items you guessed in this blank. _____ Go to your personal progress plotter following the Introduction and enter these numbers in the designated locations.

Step 3—Once again the learning begins. During the feedback step research the correct answer using Appendix A information for Examination I-2. Highlight the correct answer during your research of the reference materials. Read the entire paragraph containing the correct answer.

Helpful Hint

Follow each step carefully to realize the best return on effort. Would you consider investing your money in a venture without some chance of return on that investment? Examination preparation is no different. You are investing time expecting a significant return for that time. If, indeed, time is money then you are investing money and are due a return on that investment. Doing things right and doing the right things in examination preparation will ensure the maximum return on effort.

Examination I-2

Directions

Remove Examination I-2 from the manual. First, take a careful look at the examination. There should be 150 examination items. Notice that a blank line precedes each examination item number. This line is provided for you to enter the answer to the examination item. Write the answer **in ink**. Remember the rule about changing the answer. Our research shows that changed answers are most often changed to an incorrect answer, and, more often than not, the answer that is chosen first is correct.

If you guess the answer to a question, place an "X" or a checkmark by your answer. This step is vitally important to gain and master knowledge. We will explain how we treat the "guessed" items later in SAEP.

Take the examination. Once you complete it, go to Appendix A and score your examination. After the examination is scored, carefully follow the directions for feedback of the missed and guessed examination items.

_____ 1. Which of the following **is not** considered a function of the rescue company?
  A. Vehicle extrication
  B. Rope rescue operations
  C. Confined-space operations
  D. Stretching the initial attack line

_____ 2. The person ultimately responsible for the operations and administration of the fire department is the:
  A. mayor/supervisor.
  B. fire chief.
  C. company officer.
  D. fire fighter.

_____ 3. In all fire departments, training **must**:
  A. be a useful disciplinary tool.
  B. occur only as needed.
  C. be a continuing function.
  D. occur when time and finances allow.

_____ 4. Which of the following terms **does not** relate to fire and emergency operations?
  A. Bylaws
  B. Procedures
  C. Policies
  D. Regulations

_____ 5. What is/are the rule(s) for developing standard operating procedures (SOPs)?
  A. Firefighter safety is the first consideration for all procedures.
  B. SOPs should be brief, clear, and concise.
  C. If an SOP doesn't work, change it.
  D. All of the above

_____ 6. In the Incident Management System (IMS), the functional area that directs the organization's tactical operations to meet the strategic goals developed by command and is responsible for the management of all operations applicable to the primary mission is:
   A. Planning.
   B. Operations.
   C. Logistics.
   D. Command.

_____ 7. What is the leading cause of fire fighter injuries?
   A. Exposure to fire products
   B. Being struck by objects
   C. Overexertion and strain
   D. Exposure to chemicals

_____ 8. In the Incident Management System (IMS), the functional area responsible for all incident activities, including the development and implementation of strategic decisions, is:
   A. Planning.
   B. Operations.
   C. Logistics.
   D. Command.

_____ 9. In the Incident Management System (IMS), the functional area responsible for the collection, evaluation, dissemination, and use of information concerning the development of an incident is:
   A. Planning.
   B. Operations.
   C. Logistics.
   D. Command.

_____ 10. The picture below depicts one of the three elements of a knot or hitch. It is known as a:
   A. bight.
   B. loop.
   C. round turn.
   D. clove hitch.

_____ **11.** Elements for forming a knot are:
   A. bight, loop, and round turn.
   B. loop, bend, and crown.
   C. round turn, standing, and running.
   D. standing, bight, and hitch.

_____ **12.** The clove hitch is essentially:
   A. a half hitch.
   B. two half hitches.
   C. three loops like half hitches.
   D. a half hitch with a safety.

_____ **13.** When a rope is bent back on itself while keeping the sides parallel, a _____ has been formed.
   A. knot
   B. round turn
   C. bowline
   D. bight

_____ **14.** The combination of knots recommended to hoist a pike pole includes:
   A. a becket/sheet bend with a bight.
   B. several half hitches.
   C. bowline and half hitches.
   D. a clove hitch and half hitches.

_____ **15.** A knot well suited for joining ropes together and unlikely to slip when wet <u>best</u> describes a:
   A. becket/sheet bend.
   B. bowline knot.
   C. clove hitch.
   D. half sheep shank.

_____ **16.** Of the following four types of rope construction, which one <u>is not</u> a good choice for rescue rope?
   A. Braided
   B. Laid (twisted)
   C. Braid on braid
   D. Kernmantle

_____ **17.** The end of the rope that is used to tie a knot is called the:
   A. running end.
   B. standing part.
   C. safety end.
   D. working end.

_____ **18.** According to NFPA 1983, after being used during an actual emergency operation, life safety rope must be:
   A. inspected and put back into service for the next emergency.
   B. removed from emergency use and only used for training.
   C. removed from service and destroyed.
   D. used only as a utility rope.

_____ 19. In fire departments that have access to multiple radio channels, <u>fireground</u> operations should be:
A. on multi-channels also.
B. run by cell phone so as not to tie up the radio.
C. assigned a separate channel dedicated for use on that scene only.
D. Both A and C are correct.

_____ 20. What is the largest difference between Basic 911 and Enhanced 911?
A. Enhanced systems have the capability to provide the caller's telephone number and address.
B. Enhanced systems are used only in rural areas.
C. Basic systems are more reliable than enhanced.
D. Basic systems have the capability to provide the caller's telephone number and address.

_____ 21. Complete and accurate records should be maintained at communication centers for:
A. all responses.
B. only emergency responses.
C. only responses that may be criminally related.
D. areas of the district that generate high call volume.

_____ 22. During a fire, you hear another team call "Mayday." You should:
A. report on the radio to your supervisor advising of your location.
B. stay off the radio and listen for instructions.
C. rush into the building to find the crew calling for help.
D. activate your emergency button on your radio.

_____ 23. What are the two general types of self-contained breathing apparatus?
A. Demand and pressure-demand
B. Open-circuit and closed-circuit
C. OSHA approved and NIOSH approved
D. Compressed air and liquid oxygen

_____ 24. On breathing apparatus equipped with a low-pressure hose, the low-pressure hose brings air from the:
A. cylinder to the regulator.
B. facepiece to the exhalation valve.
C. regulator to the facepiece.
D. regulator to the high-pressure hose.

_____ 25. Limitations affecting a fire fighter's ability to use SCBA effectively are:
A. physical.
B. medical.
C. mental.
D. All of the above.

___ 26. When filling an SCBA cylinder, the cylinder must be:
   A. placed in a fragmentation containment device.
   B. placed in water.
   C. filled in the open to allow for checking of signs of weakness in the cylinder.
   D. wrapped in a blanket or towel.

___ 27. PASS is an acronym for _____ Safety System.
   A. Patient Alert
   B. Private Alert
   C. Personal Alert
   D. Passenger Alert

___ 28. An advantage of a facepiece nosecup is that it:
   A. assists in communication.
   B. helps control internal fogging.
   C. increases user time.
   D. makes breathing easier.

___ 29. Inhaled toxic gases can directly cause:
   A. disease of the lung tissue.
   B. muscle cramps in the lower extremities.
   C. blurred vision, leading to blindness.
   D. disorientation and/or amnesia.

___ 30. Which of the body's systems is most vulnerable to injury from the toxic conditions and gases encountered during firefighting operations?
   A. Circulatory
   B. Respiratory
   C. Digestive
   D. Nervous

___ 31. Manufacturers should provide users of PPE with which of the following information?
   A. Cleaning instructions
   B. MSDS information
   C. Shelf life
   D. Liability protection

___ 32. The purpose for the use of reflective trim on PPE is to:
   A. increase the visibility of the wearer to others.
   B. provide protection for material under the trim.
   C. allow the wearer to blend in with the surroundings.
   D. be more stylish than the plain PPE.

___ 33. How long does it take for the PASS device to alarm if the wearer becomes inactive?
   A. 20 seconds
   B. 30 seconds
   C. 45 seconds
   D. 60 seconds

_____ **34.** The barrier devices that <u>should</u> <u>not</u> be used at an emergency scene involving a vehicle leaking fluids is:
   A. flares.
   B. fireline tape.
   C. traffic cones.
   D. utility rope.

_____ **35.** Once overhead doors have been forced, they should be:
   A. removed.
   B. unlocked to prevent locking.
   C. locked.
   D. blocked open.

_____ **36.** One way to force a lock is to physically pull the _____ out of the door using an A-tool or a K-tool.
   A. keyhole
   B. cylinder
   C. hasp
   D. strike plate

_____ **37.** Panel, slab, and ledge are all types of _____ doors.
   A. wood swinging
   B. metal swinging
   C. overhead rolling
   D. revolving
   E. roll-up

_____ **38.** Firefighters may reasonably expect residential doors to open _____ and public building doors to open:
   A. outward, inward.
   B. inward, outward.
   C. outward, outward.
   D. inward, inward.

_____ **39.** A _____ tool provides an advantage in forcing locks, opening doors, and forcing windows.
   A. cutting
   B. striking
   C. power
   D. prying

_____ **40.** Once a fire fighter has broken a window for purposes of entry, the next action should be to:
   A. call for a charged line.
   B. carefully climb through the window.
   C. open a window on the windward side of the building.
   D. clear the entire window area of glass.

____ 41. The recommended tool for breaking a tempered plate glass window is a:
   A. battering ram.
   B. pick-head axe.
   C. flat-head axe.
   D. crowbar.

____ 42. Which of the following **is not** a correct procedure for breaking glass?
   A. Strike the top of the glass.
   B. Stand to windward side.
   C. Remove all glass particles from frame.
   D. Make sure the breaking glass is above the hands.

____ 43. When forcing entry through a wood checkrail/double-hung window where the sashes are locked at the center of the checkrail, the pry should be made at the:
   A. center of the upper sash.
   B. center of the lower sash.
   C. side of upper sash.
   D. top of lower sash.

____ 44. A simple way to force an overhead folding door is to:
   A. pry up from the bottom at both outside edges.
   B. break out a panel and operate the latch from the inside.
   C. pry open from either side at approximately waist height.
   D. drive a wedge into the bottom center.

____ 45. Opening masonry walls is often referred to as:
   A. breaching.
   B. barreling.
   C. mauling.
   D. tunneling.

____ 46. During a search of a building involved in fire, if a fire fighter becomes disoriented, the fire fighter should **attempt** to:
   A. remain calm.
   B. retrace steps to original location.
   C. seek a place of refuge and activate PASS device.
   D. All of the above.

____ 47. A _____ System helps the incident commander know who is on the fireground and where fire fighters are located.
   A. Personnel Accountability
   B. Personal Alert
   C. Personnel Attendance
   D. P.A.S.S.

____ 48. An important benefit of using a Personnel Accountability System is:
   A. knowing who is on the fireground.
   B. knowing which fire fighter has seniority.
   C. knowing which company arrived on the scene first.
   D. keeping track of which fire fighters work on which shift.

_____ 49. In the arms-length/suitcase carry of a ladder by two fire fighters, each grasps the _____, permitting the ladder to swing along side their legs at arm's length.
   A. bottom beam
   B. nearest rung
   C. inside of the bottom beam
   D. outside of the top beam

_____ 50. When a ladder is raised, it should be placed at an angle of approximately _____ to ensure a safe climb.
   A. 55°
   B. 75°
   C. 65°
   D. 45°

_____ 51. Manufacturers of fiberglass and metal ladders require that the fly section be placed:
   A. in, toward building.
   B. out, away from building.
   C. even with the window sill.
   D. either in or out, placement does not matter.

_____ 52. A <u>primary</u> safety concern when raising a ladder should be:
   A. ladder selection.
   B. teamwork and strength.
   C. possible contact with electrical wires.
   D. ladder placement and angle of inclination.

_____ 53. When lifting a ladder or other heavy object, a fire fighter should:
   A. bend at the knees and waist, lift with the legs.
   B. bend at the knees keeping the back straight, lift with the legs.
   C. bend at the knees and back, lift with the arms.
   D. keep knees straight, bend over, lift with the arms.

_____ 54. If 24 feet of a 35-foot extension ladder is needed to reach a victim, the butt of the ladder should be placed approximately _____ feet from the building.
   A. 4
   B. 6
   C. 8
   D. 11

_____ 55. When a fire fighter is to perform ventilation of a window, the ladder should be placed:
   A. even with the sill.
   B. with the ladder tip about even with the top of the window.
   C. to the leeward side.
   D. directly in front of the window with the top two rungs above the sill.

_____ 56. When a ladder is used for the rescue of an injured victim from a narrow window, the tip should be placed:
   A. alongside the window.
   B. slightly below the sill.
   C. to the windward side.
   D. to the leeward side.

_____ 57. A ladder that is selected for use in reaching the third story window or roof of a building should normally be _____ feet in length.
   A. 16 - 20
   B. 21 - 27
   C. 28 - 35
   D. 40 - 50

_____ 58. Most of the materials found in the passenger compartment of motor vehicles are:
   A. natural fibers.
   B. steel or aluminum.
   C. wood.
   D. plastic (a form of polyvinyl chloride, PVC).

_____ 59. A relatively new hazard that fire fighters must be aware of when approaching fires involving newer model vehicles is:
   A. more explosive fuels.
   B. the larger size of the vehicles.
   C. supplemental restraint systems.
   D. toxic smoke from fiberglass.

_____ 60. The smallest size hoseline the NFPA recommends for advancing on a vehicle fire is:
   A. 2_ inch.
   B. 1_ inch.
   C. 1_ inch.
   D. booster line.

_____ 61. One of the most useful tools to aid in handling a charged hoseline is a hose:
   A. wrench.
   B. jacket.
   C. strap.
   D. clamp.

_____ 62. When advancing a dry line to the point of operation, it is recommended that a fire fighter carry the nozzle by:
   A. tucking it under the arm and carrying it with the opposite hand.
   B. placing the hose over the shoulder with the nozzle in front, resting on the chest.
   C. gripping the nozzle with both hands, at waist level, keeping the nozzle in front of the fire fighter.
   D. holding the nozzle at chest level with the supply line around the waist.

_____ 63. One-inch rubber-covered and rubber-lined hose equipped with one-inch couplings is commonly called a:
  A. forestry hose.
  B. supply hose.
  C. booster hose.
  D. engine line.

_____ 64. Factors involved in most exposure fires include:
  A. ventilation.
  B. chemical chain reaction.
  C. distance.
  D. ambient temperature.

_____ 65. Which of the following **violates** a rule of personal safety?
  A. Always work in pairs or teams.
  B. Completely search one room before moving on.
  C. Remain standing even when you cannot see your feet.
  D. Before entering the building, locate more than one means of egress.

_____ 66. Firefighters conducting a search _____ if such action **will not** cause the spread of fire.
  A. may open windows to provide adequate light
  B. may open windows for ventilation
  C. should always break windows from the inside
  D. should never open windows

_____ 67. When conducting a **primary** search within a structure, a fire fighter should begin:
  A. in the center of the room.
  B. on a wall.
  C. always start with right hand pattern.
  D. under or behind furnishings.

_____ 68. When executing a blanket drag, you should:
  A. pull victim forward, place the blanket around victim, and lower the victim until flat.
  B. lift victim onto the blanket and drag feet first.
  C. roll victim on side, position blanket underneath, and roll victim back to original position.
  D. carefully work blanket under victim without moving the victim.

_____ 69. When fire fighters enter a burning building to perform rescue work, they must **first** consider:
  A. water supply.
  B. their own safety.
  C. communications.
  D. safety of victims.

___ 70. Fine water droplets and maximum high water surface area are characteristics of a stream.
   A. solid
   B. fog
   C. broken
   D. straight

___ 71. A stream designed to be as compact as possible with little shower or spray is known as a _____ stream.
   A. solid
   B. fog
   C. straight
   D. narrow-angle fog

___ 72. Nozzles with flows in excess of _____ gallons per minute **are not** recommended for handlines.
   A. 40
   B. 125
   C. 300
   D. 250

___ 73. Basement fires are difficult to fight because:
   A. cool air is rising up the stairs.
   B. fire fighters must travel down through super-heated gases and smoke.
   C. large volume streams cannot be directed from outside.
   D. they have many access points and fire fighters may not know which one to use.

___ 74. Fire stream types are generally classified as:
   A. solid and fog.
   B. straight and fog.
   C. light, medium, and heavy.
   D. direct, indirect, and combination.

___ 75. To efficiently use water during a direct attack with a solid or straight stream, the fire fighter should apply the water _____ directly on the _____ until the fire darkens down.
   A. continuously, burning fuels
   B. continuously, ceiling
   C. in short bursts, ceiling
   D. in short bursts, burning fuels

___ 76. An attack that uses the steam-generating techniques of a ceiling-level attack, along with application of the fire stream on a material burning near the floor level, is know as a(n) _____ attack.
   A. direct
   B. indirect
   C. combination
   D. blitz

77. Which of the following statements concerning automatic constant pressure nozzles **is not** correct?
   A. The nozzle automatically varies the flow rate to maintain an effective nozzle pressure.
   B. A minimum nozzle pressure is needed to maintain a good spray pattern.
   C. The nozzle person can change the flow rate by opening and closing the shut-off valve.
   D. The pump operator must change the pump setting to change the flow rate of the nozzle.

78. The standard nozzle pressure for a solid stream nozzle on a handline is _____ psi.
   A. 50
   B. 80
   C. 100
   D. 125

79. During the time a stream of water passes through space, it is influenced by velocity, _____, wind, and friction with air.
   A. absorption
   B. temperature
   C. specific density
   D. gravity

80. Nozzle controls, hydrants, valves, and hose clamps should be operated _____ to prevent a water hammer.
   A. one-half turn at a time
   B. slowly
   C. rapidly
   D. during low pressure only

81. The **primary** purpose of a spanner wrench is for use in:
   A. breaking glass.
   B. shutting-off gas valves.
   C. operating hydrant valves.
   D. tightening/loosening hose couplings.

82. A hose _____ is used to seal small cuts or breaks that may occur in fire hose or to connect mismatched or damaged couplings of the same size to stop leaking.
   A. bridge
   B. clamp
   C. jacket
   D. seal

83. One of the **best** ways to move fire hose for quick use in almost any place at ground level is the:
   A. shoulder fold carry.
   B. working line/street drag.
   C. shoulder loop carry.
   D. underarm carry.

____ 84. Devices through which water flows, used in conjunction with fire hose, such as a gate valve, are known as:
   A. apparatus.
   B. appliances.
   C. tools.
   D. flow controls.

____ 85. The standpipe connection is usually located in the _____ of a multistory building.
   A. equipment room
   B. building lobby
   C. elevator shaft (bottom)
   D. stairwell

____ 86. The two types of couplings are:
   A. male and female.
   B. brass and aluminum.
   C. threaded and nonthreaded/Storz.
   D. national standard and iron pipe.

____ 87. Combustion is the result of a _____ reaction.
   A. mechanical
   B. chemical
   C. dielectrical
   D. replenishment

____ 88. A fire in the presence of a higher-than-normal concentration of oxygen will:
   A. burn slower than normal.
   B. burn faster than normal.
   C. not be effected by the oxygen.
   D. not burn if oxygen is too rich.

____ 89. _____ is described as the point-to-point transmission of heat energy.
   A. Conduction
   B. Radiation
   C. Convection
   D. Flashover

____ 90. The chemical decomposition of a substance through the action of heat **best** defines:
   A. oxidation.
   B. pyrolysis.
   C. boiling point.
   D. heat of decompression.

____ 91. _____ is the transition between the growth and fully developed stages of fire.
   A. Flashover
   B. Backdraft
   C. Flash point
   D. Ignition temperature

____ 92. The term vapor density refers to the weight of a gas as compared to the weight of:
   A. water.
   B. air.
   C. carbon.
   D. nitrogen.

____ 93. The product of combustion that is measured in degrees of temperature to signify intensity is:
   A. heat.
   B. flame.
   C. smoke.
   D. calories.

____ 94. Pyrolysis is defined as:
   A. a chemical reaction that produces heat.
   B. the concentration level of a substance at which it will support ignition and continuous burning.
   C. a state where a balance has occurred in a mixture.
   D. decomposition or transformation of a compound caused by heat.

____ 95. Combustion is:
   A. the point at which the need for outside heat application ceases and a material sustains combustion based on its own generation of heat.
   B. a chemical reaction that liberates heat.
   C. the chemical action producing heat and light in which the heat maintains the chemical chain reaction continuing the process.
   D. the concentration level of a substance at which it will burn.

____ 96. The fire triangle is composed of:
   A. heat, chemical reaction, fuel.
   B. heat, fuel, and oxygen.
   C. oxygen, nitrogen, fuel.
   D. fuel, oxygen, LEL.

____ 97. Backdraft is:
   A. a boiling liquid/expanding vapor explosion.
   B. a layer of air that has the same temperature.
   C. the rapid ignition of smoke.
   D. flames moving through the unburned gases during a fire's progression.

____ 98. A hydrocarbon is:
   A. an ideal extinguishing agent.
   B. the basic building block of all inorganic materials.
   C. any organic compound that contains only carbon and hydrogen.
   D. a catalyst in the breakdown of molecules.

_____ **99.** Pyrolysis is associated with which of the following sources of heat?
   A. Electrical
   B. Nuclear
   C. Mechanical
   D. Chemical

_____ **100.** If a gas has a vapor density greater than one when it escapes from its container:
   A. it will rise.
   B. its movement will be dependent on wind direction and speed.
   C. its movement will be dependent on temperature.
   D. it will sink and collect at low points.

_____ **101.** Of the following tools used in ventilation operations, a _____ would be <u>best</u> for sounding the roof.
   A. 14-foot pike pole
   B. pickhead axe
   C. power saw with extended chain bar
   D. truss finder

_____ **102.** The recommended method to prevent a backdraft explosion is _____ ventilation.
   A. side
   B. lateral
   C. vertical
   D. passive

_____ **103.** When cutting through a roof, a fire fighter should attempt to:
   A. remove the ceiling joist in the ventilation hole.
   B. cut a large circular hole.
   C. make the opening square or rectangular.
   D. stand to the downwind side.

_____ **104.** When done properly, trench ventilation:
   A. will help prevent horizontal fire spread.
   B. will require the use of more water for fire suppression.
   C. consists of three separate holes cut in a U shape.
   D. will prevent the normal vertical spread of fire.

_____ **105.** An important safety precaution that should be practiced when working on a roof is to:
   A. cut all guy wires to prevent tripping over them.
   B. provide a secondary means of escape.
   C. have more than two fire fighters on the roof at all times.
   D. tie oneself to the roof ladder.

_____ **106.** Smoke and heat fills a structure starting from the:
   A. lowest point.
   B. windward side.
   C. highest point.
   D. leeward side.

___ **107.** The two types of ventilation are:
   A. natural and mechanical/forced.
   B. hydro and electric.
   C. manual and mechanical.
   D. leeward and windward.

___ **108.** At _____ °F, the length of a steel structural member will increase approximately one inch for each ten feet of total length.
   A. 800
   B. 1100
   C. 1400
   D. 1000

___ **109.** Concrete has excellent _____ strength when it cures.
   A. shear
   B. compressive
   C. torsional
   D. tensile

___ **110.** The usual cause of collapse of open web steel joist is the:
   A. amount of heat generated by the fire in a structure.
   B. poor method of construction method.
   C. impact load of fire fighters on the roof.
   D. All of the above

___ **111.** The search for and extinguishment of hidden fire and placing the building in a safe condition is known as:
   A. overhaul.
   B. secondary search.
   C. size-up.
   D. salvage.

___ **112.** When fire has burned around windows or doors, it is a good policy to:
   A. remove the entire door or window and seal with plastic.
   B. overhaul the area above the windows and doors.
   C. open the entire casing area to ensure extinguishment.
   D. go below the involved area to check for extension.

___ **113.** A fire fighter can often detect hidden fires in a concealed space by:
   A. opening up the entire concealed space.
   B. waiting for flames to appear.
   C. sight, touch, and sound.
   D. smelling for burning material.

___ **114.** Most water vacuums used by fire departments in salvage operations only have a capacity of _____ gallons.
   A. 5
   B. 2.5
   C. 7.5
   D. 10

_____ **115.** A device used to route water short distances through doors, windows, or other openings is a:
   A. water chute.
   B. carryall.
   C. floor runner.
   D. catchall.

_____ **116.** When making a water chute using a ladder and salvage covers, what other item(s) is/are required?
   A. Halligan tool
   B. Pike poles
   C. Utility rope
   D. Hose line

_____ **117.** The basic premise of salvage operations is:
   A. to prevent fire extension.
   B. to protect fire department property from being damaged at the fire scene.
   C. to separate or protect interior and exterior materials from the harmful environment.
   D. to provide better information to the fire inspector.

_____ **118.** If salvage operations are going on while active suppression operations are taking place, how should the salvage crew be dressed?
   A. Station work uniforms are acceptable.
   B. They should wear full protective clothing, including SCBA.
   C. They should wear full protective clothing less SCBA.
   D. Gloves and a helmet with eyeshield are appropriate attire.

_____ **119.** The shoulder toss is done by a single fire fighter and is used to cover:
   A. small fragile items.
   B. fragile items a little taller than the fire fighter.
   C. items a little taller than the fire fighter.
   D. large unbreakable items (i.e., rack storage).

_____ **120.** When the water flow alarm (water gong) sounds, this indicates that:
   A. water has stopped flowing in the system.
   B. heat detection devices have been activated and one may expect the deluge set to begin discharging water momentarily.
   C. water is flowing in the system.
   D. a heat actuating device has been activated and someone should turn the main sprinkler valve to the open position.

_____ **121.** Portable water tanks should be positioned in a location that allows easy access from:
   A. multiple directions.
   B. only one direction.
   C. the windward side.
   D. the leeward side.

____122. Water supply is one of the most critical elements of firefighting because:
  A. of the great expense in obtaining it.
  B. water is the most common extinguishing agent.
  C. water freezes at high temperatures.
  D. of its ability to suffocate a fire.

____123. Available flow/static pressure is the:
  A. rate and quantity of water delivered.
  B. amount of water flowing from the discharge side of the pump.
  C. amount of water required to put out the fire.
  D. amount of water that can be moved to put out the fire.

____124. Tenders combined with _____ can efficiently provide large volumes of water to a fire ground operation.
  A. large diameter hose
  B. automatic nozzles
  C. portable water tanks
  D. ladder trucks

____125. The **most common** water distribution system is a _____ system.
  A. pumped
  B. combination pumped/gravity
  C. gravity
  D. tender shuttle

____126. The two major hydrant types are:
  A. wet barrel and dry barrel.
  B. high-pressure and low-pressure.
  C. ground water and surface water.
  D. treated water and untreated water.

____127. A Class D fire involves:
  A. combustible metals.
  B. flammable liquids.
  C. electrical equipment.
  D. ordinary combustibles.

____128. Fires involving flammable liquids, greases, and gases where the smothering or blanketing effect is needed are _____ fires.
  A. Class A
  B. Class B
  C. Class C
  D. Class D

____129. Extinguishers suitable for Class D fires can be identified by a _____ containing the letter D.
  A. blue circle
  B. yellow star
  C. green triangle
  D. red square

130. A carbon dioxide (CO_2) extinguisher's means of discharge is:
  A. chemical reaction.
  B. stored liquefied compressed gas.
  C. cartridge activation.
  D. manual hand-pump.

131. A pump tank extinguisher rated as 4-A can be expected to extinguish approximately _____ as much fire as one rated 2-A.
  A. twice
  B. three times
  C. four times
  D. eight times

132. A fire extinguisher bearing the symbols shown below would be suitable for extinguishing _____ fires.
  A. Class A, B, and C
  B. Class B and C
  C. Class A and B
  D. Class A and C

133. Class K fires involve:
  A. atomic material.
  B. computer network equipment.
  C. hazardous waste.
  D. high temperature cooking oils, such as vegetable or animal oils and fats.

134. The **safest** way for a fire fighter to disconnect electrical service to a building is to:
  A. remove the electrical meter.
  B. cut the service entrance wires.
  C. shut off the main breakers/switch at the service panel.
  D. pull the main breaker at the power pole.

135. Aspect is the:
  A. direction a slope faces.
  B. measure of the steepness of a slope.
  C. measure of the roughness of a slope.
  D. measure of the direction in which the wind moves across a slope.

136. The **most important** factors that affect wildland firefighting are:
  A. fuel, equipment, and location.
  B. topography, resources, and time of day.
  C. fuel, weather, and topography.
  D. staffing, resources, and apparatus.

____137. In addition to providing a service, private dwelling inspections also:
   A. afford the opportunity to introduce members of the fire department.
   B. provide an educational and advisory service.
   C. afford an opportunity to impress upon the public the benefits of the fire department.
   D. All of the above.

____138. Which of the following statements regarding preparing for an inspection visit is <u>incorrect</u>?
   A. Plan the area to be inspected.
   B. Review occupancy files prior to leaving the station.
   C. Inspections are performed around firefighting schedules and not the schedule of the business owner.
   D. Give consideration to the type of activities conducted at the business relative to the time of day chosen for the inspection.

____139. With regard to portable fire extinguishers, which of the following situations would a fire fighter bring to the occupant/owner's attention?
   A. The extinguisher has the proper classification and rating for its location.
   B. The pressure gauge indicates that the extinguisher isn't properly charged.
   C. The fire extinguisher is hung on the wall and is easily visible and accessible.
   D. The tag indicates that the extinguisher has been serviced to local laws.

____140. Which of the following is a factor in deciding to preplan a structure or area?
   A. Type of hazards expected
   B. Complexity of firefighting operations
   C. Nature of activities conducted at the occupancy
   D. All of the above.

____141. A drawing or diagram of a building or area as seen from directly overhead is the definition of a _____ view/sketch.
   A. sectional
   B. plan
   C. good
   D. administrative

____142. Fire departments should educate _____ to recognize potential hazards and take appropriate corrective action.
   A. preschoolers
   B. the elderly
   C. adults
   D. citizens of all ages

____143. Defective SCBA cylinder units should be:
   A. repaired by the person who discovers the defect.
   B. removed from service.
   C. put on reserve fire apparatus.
   D. filled to 80% capacity.

____144. The _____ must be stamped or labeled on a compressed air cylinder.
   A. fire department's initials
   B. last hydrostatic test date
   C. date on which the cylinder must be hydrostatically tested
   D. last fill date

____145. Worn, damaged, and deteriorated parts of a SCBA must be replaced according to:
   A. past practice.
   B. NIOSH/OSHA Respiratory Protection Act.
   C. manufacturer's instructions.
   D. the wearer's recommendations.

____146. Composite SCBA cylinders must be hydrostatically tested every:
   A. year.
   B. three years.
   C. five years.
   D. ten years.

____147. It is permissible to use paint on a fire department ladder:
   A. to mark the bed section on multisection ladders.
   B. when there is a possibility of dry rot.
   C. to mark the ladder ends for visibility.
   D. when salt water may be a problem.

____148. Which of the following **is not** a method for preventing mechanical damage to fire hose?
   A. Avoid closing the nozzle abruptly.
   B. Remove wet hose from apparatus and replace with dry hose.
   C. Prevent vehicles from driving over fire hose.
   D. Avoid laying hose over rough, sharp edges or corners.

____149. Of the following, the most important factor relating to the life of a hose is the:
   A. method of service testing.
   B. manufacturing process.
   C. care it receives.
   D. number of jackets used in construction.

_____**150.** In the care and maintenance of fire hose, which one of the following procedures **is not** a good practice?
   A. Dry hose in a hose dryer.
   B. Keep the exterior of woven jacketed hose dry.
   C. Allow hose to remain in a heated area after it dries.
   D. Use moderate temperatures for drying.

---- Helpful Hint ----
Try to determine why you selected the wrong answer. Usually something influenced your selection. Focus on the difference between your wrong answer and the right answer. Carefully read and study the entire paragraph containing the correct answer. Highlight the answer just as you did for Examination I-1.

Did you score higher than 80% on Examination I-2? Circle Yes or No **in ink**. (We will return to your Yes or No answer to this question later in SAEP).

Examination I-3, Confirming What You Mastered

During Examination I-3, progress will be made in reinforcing what you have learned and improving your examination-taking skills. This examination contains approximately 70% of the examination items you have already answered and several new examination items. Follow the steps carefully to realize the best return on effort.

Step 1—Take Examination I-3. When you have completed Examination I-3, go to Appendix A and compare your answers with the correct answers.

Step 2—Score Examination I-3. How many examination items did you miss? Write the number of missed examination items in the blank <u>in ink</u>. _____ Enter the number of examination items you guessed in this blank. _____ Go to your personal progress plotter following the Introduction and enter these numbers in the designated locations.

Step 3—During the feedback step, research the correct answer using Appendix A information for Examination I-3. Highlight the correct answer during your research of the reference materials. Read the entire paragraph containing the correct answer.

Examination I-3

Directions

Remove Examination I-3 from the manual. First, take a careful look at the examination. There should be 200 examination items. Notice that a blank line precedes each examination item number. This line is provided for you to enter the answer to the examination item. Write the answer **in ink**. Remember the rule about changing the answer. Our research shows that changed answers are most often changed to an incorrect answer, and more often than not the answer that is chosen first is correct.

If you guess the answer to a question, place an "X" or a checkmark by your answer. This step is vitally important to gain and master knowledge. We will explain how we treat the "guessed" items later in SAEP.

Take the examination. Once you complete it, go to Appendix A and score your examination. Once the examination is scored, carefully follow the directions for feedback of the missed and guessed examination items.

_____ **1.** What is/are the rule(s) for developing standard operating procedures (SOPs)?
   A. Firefighter safety is the first consideration for all procedures.
   B. SOPs should be brief, clear, and concise.
   C. If an SOP doesn't work, change it.
   D. All of the above

_____ **2.** Which of the following items **is not** important when factored into the fire fighter's safety equation?
   A. Gender
   B. Attitude
   C. Training
   D. Fitness/health

_____ **3.** Making equipment safe is addressed in what way?
   A. Selection
   B. Inspection and maintenance
   C. Application
   D. All of the above

_____ **4.** The chain of command.
   A. includes training, safety, finance, and logistics.
   B. includes incident stabilization, life safety, and property conservation.
   C. allows for supervision of five people by one person.
   D. is the pathway of responsibility from the highest level of the department to the lowest.

_____ **5.** A fireground management system that addresses procedures for controlling personnel, facilities, equipment, and communications and is designed to be expanded as needed is the:
   A. Critical Incident Management System.
   B. Emergency Management System.
   C. Task Management System.
   D. Incident Management System.

_____ 6. The **primary** use of the _____ is to attach a rope to a cylindrical object.
   A. clove hitch
   B. bowline
   C. becket/sheet bend
   D. rescue knot

_____ 7. The part of the rope that is used for work such as hoisting or pulling is called the:
   A. working end.
   B. round turn.
   C. running end.
   D. standing part.

_____ 8. When hoisting a ladder, the rope should be threaded and secured through the ladder _____ the distance from the top.
   A. 1/6
   B. 1/5
   C. 1/3
   D. 1/2

_____ 9. The picture below depicts one of the three elements of a knot or hitch. It is known as a:
   A. bight.
   B. loop.
   C. round turn.
   D. clove hitch.

_____ 10. The _____ is tied as shown below.
   A. bowline
   B. half-hitch
   C. clove hitch
   D. figure-eight

_____ 11. The <u>first</u> <u>step</u> in tying the becket/sheet bend is to form a:
   A. loop.
   B. bight.
   C. round turn.
   D. half hitch.

_____ 12. A _____ is a double section of rope, usually made along the standing part, that forms a U-turn in the rope that does not cross itself.
   A. loop
   B. dressing
   C. bight
   D. knot

_____ 13. Which of the following statements regarding the washing of synthetic fiber ropes is <u>incorrect</u>?
   A. Use only cold tap water.
   B. Use bleach to remove grease or oils.
   C. When using a washing machine, use only a front-loading machine (with a glass window) without an agitator.
   D. Kernmantle rope can be placed in a mesh bag or "chained" and washed in a front-loading washing machine.

_____ 14. A pike pole should:
   A. be hoisted point down.
   B. be hoisted sideways.
   C. be hoisted point up.
   D. not be hoisted due to the risk of injury to fire fighters on the ground.

_____ **15.** Ropes should be inspected:
   A. after every use.
   B. monthly.
   C. annually.
   D. prior to being used.

_____ **16.** All fire department radio operations must follow rules of the:
   A. NFPA.
   B. DOT.
   C. FCC.
   D. CFR.

_____ **17.** Which of the following is not identified as an alerting system for a staffed fire station?
   A. Vocal alarm
   B. Fax machine
   C. Computerized line printer
   D. Radio with alert tone

_____ **18.** All of the following are considered valuable characteristics or traits for a person who receives emergency calls except:
   A. the ability to perform multiple tasks.
   B. an inability to retain composure.
   C. the ability to remember details and recall information easily.
   D. the ability to exercise voice control.

_____ **19.** Which of the following is not a proper radio procedure for fire personnel?
   A. Transmit when the airwaves are clear.
   B. Hold the radio/microphone one to two inches from the mouth at a 45° angle.
   C. Speak as you key the microphone to save time.
   D. Think about what is going to be said prior to transmitting.

_____ **20.** The bypass valve on a self-contained breathing apparatus is used:
   A. during hazardous materials incidents.
   B. in emergency situations involving a malfunctioning regulator.
   C. to clear the mask of unwanted condensation.
   D. to cool the face piece when high heat is encountered.

_____ **21.** When tightening the straps on a SCBA facepiece, the _____ straps should always be tightened first.
   A. lower
   B. temple
   C. upper
   D. harness

_____ **22.** As the oxygen supply in any given area falls below _____ percent, unconsciousness can occur.
   A. 21
   B. 17
   C. 12
   D. 9

23. Rescue from sewers, grain silos, and similarly confined spaces requires the use of self-contained breathing apparatus due to the danger of:
A. toxic gases.
B. oxygen deficiency.
C. ambient temperature.
D. Both A and B are correct.

24. PASS devices are designed to assist rescuers attempting to:
A. move through traffic while responding to an incident.
B. locate trapped fire fighters.
C. eject smoke from a building.
D. roll hose faster than by hand.

25. The proper position of the bypass valve on positive-pressure SCBA under normal conditions is:
A. fully open.
B. cracked open.
C. fully closed.
D. open three full turns.

26. On breathing apparatus equipped with a low-pressure hose, the low-pressure hose brings air from the:
A. cylinder to the regulator.
B. facepiece to the exhalation valve.
C. regulator to the facepiece.
D. regulator to the high-pressure hose.

27. Which of the following **is not** a component of an open-circuit breathing apparatus?
A. Regulator
B. Facepiece
C. Low-pressure alarm
D. Oxygen cylinder

28. An advantage of a facepiece nosecup is that it:
A. assists in communication.
B. helps control internal fogging.
C. increases user time.
D. makes breathing easier.

29. Which of the body's systems is most vulnerable to injury from the toxic conditions and gases encountered during firefighting operations?
A. Circulatory
B. Respiratory
C. Digestive
D. Nervous

30. Atmospheres are classified as oxygen deficient when they fall below _____ percent oxygen.
   A. 25
   B. 19.5
   C. 16
   D. 13.5

31. Toxic substances found in smoke include:
   A. carbon monoxide.
   B. carbon dioxide.
   C. tar.
   D. All of the above

32. Which of the following **is not** a limitation of an SCBA unit based on design and size?
   A. Peripheral vision is reduced.
   B. The quantity of air is immeasurable.
   C. The weight and bulk of the unit restricts mobility and agility for fire fighters.
   D. Ability to communicate is decreased.

33. SCBA limitations that the fire fighter/user should be aware of is/are:
   A. increased physical stress may cause anxiety.
   B. SCBA and PPE add approximately 40-50 pounds of weight to the fire fighter.
   C. the degree of training or experience users have with SCBA effects their level of self-confidence.
   D. All of the above.

34. What type of SCBA is **most commonly** used in the fire service?
   A. Closed-circuit
   B. Open-circuit
   C. Multicircuit
   D. Rebreather

35. What is the **first step** a fire fighter should take if their SCBA becomes damaged or malfunctions?
   A. Use the protective hood as a filter.
   B. Manually activate the PASS device.
   C. Remain calm and rely on previous training.
   D. Run for the nearest exit as quickly as possible.

36. Firefighters should practice reduced profile maneuvers with SCBA, but they must realize that this procedure:
   A. should be used at every fire.
   B. should be used as a last resort for an emergency escape from a hostile environment.
   C. does not require much practice to attain proficiency.
   D. is best left to the chief officers.

_____ 37. Which of the following **is not** a safety practice when riding on an apparatus?
   A. Riding in a fully enclosed cab
   B. Dressing while the apparatus is in motion
   C. Being seated with seat belts fastened
   D. Not standing on the apparatus

_____ 38. Which of the following **is not** one of the components of the layered protection system in structural PPE?
   A. Vapor/moisture barrier
   B. Waterproof outer layer
   C. Thermal barrier
   D. Outer shell

_____ 39. The purpose for the use of reflective trim on PPE is to:
   A. increase the visibility of the wearer to others.
   B. provide protection for material under the trim.
   C. allow the wearer to blend in with the surroundings.
   D. be more stylish than the plain PPE.

_____ 40. Which of the following statements regarding a work uniform is **incorrect**?
   A. NFPA 1975 standard outlines manufacturers' guidelines and requirements.
   B. Uniforms add protective measure to fire fighters engaged in support activities.
   C. A uniform that meets NFPA standards is designed to protect the wearer in IDLH atmospheres.
   D. A work uniform that meets NFPA standards is not designed to protect the wearer from IDLH atmospheres.

_____ 41. Which of the following statements is **incorrect**?
   A. Goggles, safety glasses, and wrap-around shields are forms of eye protection.
   B. Proximity PPE shields are coated to create a mirrored surface.
   C. An SCBA face shield can provide primary eye protection.
   D. A structural helmet face shield can provide primary eye protection.

_____ 42. How long does it take for the PASS device to alarm if the wearer becomes inactive?
   A. 20 seconds
   B. 30 seconds
   C. 45 seconds
   D. 60 seconds

_____ 43. Tempered plate glass should be broken only as a last resort. It is recommended that it be shattered by striking:
   A. in the center with a large, blunt object.
   B. at the lowest corner with the pick end of a fire axe.
   C. with a flat-headed axe with hands above the head.
   D. with a flat-headed axe at the highest corner.

___ **44.** In what type of occupancy will fire fighters usually find sliding doors?
   A. Barns or warehouses
   B. Commercial
   C. Residential
   D. Institutional

___ **45.** The <u>primary</u> rule of forcible entry is:
   A. "Look before you leap."
   B. "Try before you pry."
   C. "Always stand to leeward."
   D. "Sharp edge pointed down and ahead."

___ **46.** One way to force a lock is to physically pull the _____ out of the door using an A-tool or a K-tool.
   A. keyhole
   B. cylinder
   C. hasp
   D. strike plate

___ **47.** Of the choices listed below, the <u>least desirable</u> point to force entry into a structure is:
   A. a check rail window.
   B. a swinging door.
   C. plate glass.
   D. a revolving door.

___ **48.** Firefighters may reasonably expect residential doors to open _____ and public building doors to open:
   A. outward, inward.
   B. inward, outward.
   C. outward, outward.
   D. inward, inward.

___ **49.** When forcing entry through a wood checkrail/double-hung window where the sashes are locked at the center of the checkrail, the pry should be made at the:
   A. center of the upper sash.
   B. center of the lower sash.
   C. side of upper sash.
   D. top of lower sash.

___ **50.** The <u>best</u> and most practical way to gain entry through a casement window is to:
   A. use an axe or other tool to break out the cross members.
   B. pry at the midpoint between the upper and lower windows to force the lock.
   C. pry the bottom section upward from a center point to break the lock.
   D. break out one of the lowest panes of glass and reach through to disengage the lock.

____ **51.** A simple way to force an overhead folding door is to:
　　A. pry up from the bottom at both outside edges.
　　B. break out a panel and operate the latch from the inside.
　　C. pry open from either side at approximately waist height.
　　D. drive a wedge into the bottom center.

____ **52.** Fire axes should be carried:
　　A. over the shoulder.
　　B. by the handle with the blade hanging close to the ground.
　　C. by the handle with the blade toward the body.
　　D. with the axe blade pointed away from the body and the pick-end shielded.

____ **53.** Forcible entry is:
　　A. forcing less confident fire fighters into the structure.
　　B. the ability to gain access to unsecured buildings.
　　C. the ability to gain entry to secured areas and buildings at fires and other operations.
　　D. forcing openings in a structure to facilitate effective ventilation.

____ **54.** Which of the following tools <u>does not</u> belong to the cutting group of tools?
　　A. Saws
　　B. Bolt cutters
　　C. Center punch
　　D. Torches

____ **55.** Power saws are available in two basic types, the chain saw and the:
　　A. band saw.
　　B. hand saw.
　　C. timber saw.
　　D. rotary saw with circular blade.

____ **56.** A cutting torch has a flame temperature in excess of:
　　A. 10,000°F.
　　B. 15,000°F.
　　C. 12,000°F.
　　D. 5,000°F.

____ **57.** Which of the following statements regarding power saw safety is <u>incorrect</u>?
　　A. Use the right blade for the material being cut.
　　B. The saw should be started on level ground and carried up the ladder while running.
　　C. Power saws require two fire fighters-the saw operator and a guide fire fighter.
　　D. Conduct daily checks for operation and blade condition.

____ **58.** _____ glass is also known as safety glass.
　　A. Regular/plate
　　B. Laminated
　　C. Tempered
　　D. Wire

_____ **59.** Because of the high risks associated with search and rescue:
   A. all floors must be searched simultaneously.
   B. it is done by teams of fire fighters.
   C. one fire fighter should accomplish the task.
   D. it is not recommended in large structures.

_____ **60.** Safety requires that fire fighters using self-contained breathing apparatus must work:
   A. alone.
   B. in contained areas.
   C. in pairs.
   D. with a lifeline.

_____ **61.** Before fire fighters enter a burning building to perform rescue work, they must first consider:
   A. manpower on the scene.
   B. weather conditions.
   C. damaging evidence of forced entry.
   D. their own safety.

_____ **62.** During a search of a building involved in fire, if a fire fighter becomes disoriented, the fire fighter should <u>attempt</u> to:
   A. remain calm.
   B. retrace steps to original location.
   C. seek a place of refuge and activate PASS device.
   D. All of the above.

_____ **63.** One purpose of a Personnel Accountability System is to:
   A. keep track of fire fighters' salaries.
   B. identify trapped or injured fire fighters.
   C. create a job complaint forum.
   D. help the Fire Chief control sick leave abuse.

_____ **64.** Using a ladder at a reduced angle (less than 75°) drastically reduces the _____ of the ladder and _____ the possibility of slippage.
   A. weight, reduces
   B. load-carrying capacity, increases
   C. curvature, increases
   D. width, reduces

_____ **65.** In the arms-length/suitcase carry of a ladder by two fire fighters, each grasps the _____, permitting the ladder to swing along side their legs at arm's length.
   A. bottom beam
   B. nearest rung
   C. inside of the bottom beam
   D. outside of the top beam

_____ 66. Manufacturers of fiberglass and metal ladders require that the fly section be placed:
   A. in, toward building.
   B. out, away from building.
   C. even with the window sill.
   D. either in or out, placement does not matter.

_____ 67. When assigned to raise a ladder to a second story roof, a fire fighter should choose a _____ foot ladder.
   A. 16 - 20
   B. 40 - 50
   C. 28 - 35
   D. 20 - 24

_____ 68. When a fire fighter is to perform ventilation of a window, the ladder should be placed:
   A. even with the sill.
   B. with the ladder tip about even with the top of the window.
   C. to the leeward side.
   D. directly in front of the window with the top two rungs above the sill.

_____ 69. Other than a ladder safety belt, a fire fighter can be safely secured to a ground ladder using:
   A. a rope.
   B. an arm lock.
   C. hose strap.
   D. a leg lock.

_____ 70. A ladder that is selected for use in reaching the third story window or roof of a building should normally be _____ feet in length.
   A. 16 - 20
   B. 21 - 27
   C. 28 - 35
   D. 40 - 50

_____ 71. When a fire fighter is carrying a ladder, all lifting and lowering should be done using the _____ muscles.
   A. back
   B. wrist
   C. leg
   D. upper arm

_____ 72. The lower section of an extension ladder is known as the:
   A. fly.
   B. base or bed.
   C. beam.
   D. truss.

_____ 73. During a two-fire fighter raise of a single ladder, it is the responsibility of the _____ to determine the proper placement distance from the building.
   A. driver/operator
   B. fire fighter at the butt (heeler)
   C. company officer
   D. incident commander

_____ 74. While working from a ground ladder, the hook on the fire fighter's safety harness should be:
   A. attached to a rung.
   B. attached to a beam.
   C. secured to a lifeline.
   D. attached to the halyard.

_____ 75. While a roof ladder is being taken up an extension ladder, the hooks are:
   A. retracted or bedded.
   B. extended toward the fire fighters as handles.
   C. extended outward from the fire fighters.
   D. on the lower end of the ladder.

_____ 76. What is the purpose of a sensor label on fire department ladders?
   A. The label is used to warn fire fighters that the ladder has been dropped and may have suffered structural damage.
   B. The heat sensitive label is affixed to the ladder to alert fire fighters that the ladder has been exposed to potentially damaging heat levels.
   C. The temperature sensitive label is affixed to the ladder to alert fire fighters that the ladder has been exposed to subfreezing temperatures that may have structurally damaged the ladder.
   D. The label is used to warn fire fighters that the ladder is placed on an angle that makes working from it unsafe.

_____ 77. Which of the following is not considered one of the basic tasks of the ladder company crew?
   A. Advancing hose lines to extinguish the fire
   B. Forcible entry to gain access to the fire
   C. Gaining access to the roof for ventilation
   D. Gaining access to upper floors for search and rescue

_____ 78. When an extension ladder can be raised more than _____ feet, it must be equipped with staypoles.
   A. 35
   B. 32
   C. 40
   D. 50

_____ 79. The proper distance the foot or butt of a ladder should stand out from a building is _____ of the working distance of the ladder from the base of the wall.
   A. one-half
   B. one-quarter
   C. one-third
   D. one-eighth

_____ 80. Which of the following hoselines **should not** be used on vehicle fires?
   A. Booster hose lines
   B. 1½" attack lines
   C. 2½" attack lines
   D. 1¾" lines

_____ 81. Most of the materials found in the passenger compartment of motor vehicles are:
   A. natural fibers.
   B. steel or aluminum.
   C. wood.
   D. plastic (a form of polyvinyl chloride, PVC).

_____ 82. During search operations, the fire fighter should occasionally pause and:
   A. rest to conserve air.
   B. regroup with all those searching.
   C. listen for calls or signals for help.
   D. leave a trail to aid escape.

_____ 83. When conducting a **primary** search within a structure, a fire fighter should begin:
   A. in the center of the room.
   B. on a wall.
   C. always start with right hand pattern.
   D. under or behind furnishings.

_____ 84. When fire fighters enter a burning building to perform rescue work, they must **first** consider:
   A. water supply.
   B. their own safety.
   C. communications.
   D. safety of victims.

_____ 85. Firefighters use the "right hand"/"left hand" search in _____ occupancies.
   A. mercantile
   B. industrial
   C. residential
   D. All of the above

_____ 86. Which of the following statements is <u>incorrect</u>?
   A. The secondary search is the most dangerous.
   B. Searching a building is completed in two different operations-primary and secondary search.
   C. During primary search, the team is often ahead of attack lines and may be above the fire.
   D. The primary search takes place in a rapid but thorough manner in areas most likely to have victims.

_____ 87. When lifting an object during a search, fire fighters should <u>always</u>:
   A. use their back to lift.
   B. use their legs to lift, not their back.
   C. try to twist and reach at the same time.
   D. lift with their arms and back.

_____ 88. Which of the following statements is <u>incorrect</u>?
   A. Use care when dragging a fire fighter wearing an SCBA to prevent breaking the seal.
   B. All types of rescue drags provide good spinal immobilization.
   C. Drags are intended to be used where greater harm will come to the patient if he is not immediately moved.
   D. Rescue drags do not provide effective spinal immobilization.

_____ 89. Which of the following can indicate the possibility of a building collapse?
   A. Loose mortar joints
   B. Leaning of exterior walls
   C. Creaks and cracking sounds
   D. All of the above

_____ 90. Solid-stream handlines are designed to be operated at a nozzle pressure of _____ psi.
   A. 50
   B. 75
   C. 90
   D. 100

_____ 91. When a hose is being rolled into twin rolls secured together by a portion of hose itself, it is called a _____ roll.
   A. donut
   B. hi-rise
   C. self-locking
   D. straight

_____ 92. Prior to entering a fire area with a charged hoseline, the nozzle operator should first:
   A. bleed air from the line.
   B. wait for a building layout.
   C. wait for the power to be shut off.
   D. wait for direction from the pump operator.

___ 93. When water flowing through a fire hose or pipe is suddenly stopped, the resulting surge is referred to as:
A. static energy absorption.
B. a water hammer.
C. flow pressure.
D. residual pressure.

___ 94. A fire stream used in an indirect attack in a structure fire should be shut down before:
A. all spot fires are extinguished.
B. a backup line becomes available.
C. ventilation occurs.
D. thermal layering is disturbed.

___ 95. Fine water droplets and maximum high water surface area are characteristics of a _____ stream.
A. solid
B. fog
C. broken
D. straight

___ 96. Nozzles with flows in excess of _____ gallons per minute **are not** recommended for handlines.
A. 40
B. 125
C. 300
D. 250

___ 97. Fire stream types are generally classified as:
A. solid and fog.
B. straight and fog.
C. light, medium, and heavy.
D. direct, indirect, and combination.

___ 98. To efficiently use water during a direct attack with a solid or straight stream, the fire fighter should apply the water _____ directly on the _____ until the fire darkens down.
A. continuously, burning fuels
B. continuously, ceiling
C. in short bursts, ceiling
D. in short bursts, burning fuels

___ 99. An attack that uses the steam-generating techniques of a ceiling-level attack, along with application of the fire stream on a material burning near the floor level, is know as a(n) _____ attack.
A. direct
B. indirect
C. combination
D. blitz

_____100. Which of the following statements concerning automatic constant pressure nozzles **is** **not** correct?
   A. The nozzle automatically varies the flow rate to maintain an effective nozzle pressure.
   B. A minimum nozzle pressure is needed to maintain a good spray pattern.
   C. The nozzle person can change the flow rate by opening and closing the shut-off valve.
   D. The pump operator must change the pump setting to change the flow rate of the nozzle.

_____101. When advancing a **dry** hoseline up a ladder, fire fighters should position themselves on the ladder:
   A. on opposing sides of the beam.
   B. no more than six feet apart.
   C. with no more than one fire fighter per section
   D. within arms' reach of each other

_____102. When advancing a hoseline into a burning structure, it is recommended to bleed the air out of the line _____ entering the fire area.
   A. immediately after
   B. before
   C. soon after
   D. upon

_____103. The **primary** purpose of a spanner wrench is for use in:
   A. breaking glass.
   B. shutting-off gas valves.
   C. operating hydrant valves.
   D. tightening/loosening hose couplings.

_____104. A hose _____ is used to seal small cuts or breaks that may occur in fire hose or to connect mismatched or damaged couplings of the same size to stop leaking.
   A. bridge
   B. clamp
   C. jacket
   D. seal

_____105. The standpipe connection is usually located in the _____ of a multistory building.
   A. equipment room
   B. building lobby
   C. elevator shaft (bottom)
   D. stairwell

_____106. The _____ method is performed with **two** fire fighters and used for breaking tight couplings without the use of a spanner wrench.
   A. stiff-arm
   B. foot-tilt
   C. knee-press
   D. coupling-tilt

____107. A water hammer is:
   A. a hydraulic tool used for breaking apart concrete.
   B. the pressure surge that occurs when water valves or nozzles are slowly opened and closed.
   C. the use of water to assist in overhaul by washing pieces of plaster and other building materials away from the structure.
   D. the pressure surge that occurs when water valves or nozzles are rapidly opened or closed.

____108. A dutchman is:
   A. used when loading hose and has a short fold or a reverse fold with a coupling at the point where a fold should occur.
   B. a process for uncoupling hose.
   C. a short length of hose that connects a deluge set to the engine company's pump.
   D. the fold in a twin donut roll.

____109. Which of the following statements regarding the accordion load is **incorrect**?
   A. It is easy to load and unload.
   B. It can be used as an additional supply line.
   C. It is ideal for making shoulder loads.
   D. It is the preferred way to load large diameter hose (LDH).

____110. Which of the following methods of fire attack utilizes flowing water at the base of the fire?
   A. Direct
   B. Indirect
   C. Combination
   D. Offensive

____111. Which of the following **best** describes a recommended procedure for horizontal ventilation of a building?
   A. Open the lower windows on the leeward side first, allowing heated gases to escape; then open the upper windows on the windward side.
   B. Open the upper windows on the leeward side first, allowing heated gases to escape; then open the lower windows on the windward side.
   C. Open the lower windows on the windward side first, allowing fresh air to enter; then open the upper windows on the leeward side.
   D. Open the upper windows on the windward side first, allowing fresh air to enter; then open the lower windows on the leeward side.

____112. The side of a building that the wind is striking is called the _____ side. The opposite side is called the _____ side.
   A. windward, leeward
   B. leeward, windward
   C. downwind, upwind
   D. upwind, downwind

___ 113. When ventilating a building, all of the following procedures can be used by fire fighters <u>except</u> _____ ventilation.
  A. vertical
  B. horizontal
  C. forced
  D. manual

___ 114. The <u>primary</u> function of smoke ejectors or exhaust fans is:
  A. localizing the fire.
  B. removing heat and smoke.
  C. providing fresh air for attack crews.
  D. removing lighter-than-air gases.

___ 115. Forced/mechanical ventilation is accomplished by blowers, fans, or:
  A. removal of windows.
  B. vertical openings.
  C. fog streams.
  D. natural wind currents.

___ 116. The phase of fire characterized by temperature decline and diminishing fire is called:
  A. ignition.
  B. growth.
  C. fully developed.
  D. decay.

___ 117. When the colder of two bodies in contact absorbs heat until both objects are the same temperature, _____ has taken place.
  A. specific heat equalization
  B. heat transfer
  C. latent heat vaporization
  D. osmosis transfer

___ 118. The principle which <u>most closely</u> describes how water extinguishes fire is:
  A. removal of fuel.
  B. reduction of temperature.
  C. exclusion of oxygen.
  D. inhibition of chain reaction.

___ 119. A fire in the presence of a higher-than-normal concentration of oxygen will:
  A. burn slower than normal.
  B. burn faster than normal.
  C. not be effected by the oxygen.
  D. not burn if oxygen is too rich.

___ 120. _____ is described as the point-to-point transmission of heat energy.
  A. Conduction
  B. Radiation
  C. Convection
  D. Flashover

_____ 121. The term vapor density refers to the weight of a gas as compared to the weight of:
   A. water.
   B. air.
   C. carbon.
   D. nitrogen.

_____ 122. A product of combustion that contains a mixture of oxygen, hydrogen cyanide, CO_2, CO, and finely divided carbon particles is:
   A. heat.
   B. flame.
   C. smoke.
   D. vapor.

_____ 123. The movement of heat through a steel beam to an unexposed part of a building where it starts another fire is an example of:
   A. conduction.
   B. radiation.
   C. convection.
   D. direct-flame contact.

_____ 124. The _____ stage occurs when all combustible materials in the compartment are involved in the fire.
   A. ignition
   B. growth
   C. flashover
   D. fully-developed

_____ 125. A hydrocarbon is:
   A. an ideal extinguishing agent.
   B. the basic building block of all inorganic materials.
   C. any organic compound that contains only carbon and hydrogen.
   D. a catalyst in the breakdown of molecules.

_____ 126. If a gas has a vapor density greater than one when it escapes from its container:
   A. it will rise.
   B. its movement will be dependent on wind direction and speed.
   C. its movement will be dependent on temperature.
   D. it will sink and collect at low points.

_____ 127. Ignition is:
   A. the term used to denote a place where heat is drained away from a source.
   B. the ability of a material to sustain combustion.
   C. a catalyst in the breakdown of molecules.
   D. a chemical reaction that absorbs heat.

_____ 128. Which type of heat transfer is a major contributor to flashover?
   A. Radiation
   B. Conduction
   C. Convection
   D. Nuclear

129. Of the following tools used in ventilation operations, a _____ would be **best** for sounding the roof.
   A. 14-foot pike pole
   B. pickhead axe
   C. power saw with extended chain bar
   D. truss finder

130. When done properly, trench ventilation:
   A. will help prevent horizontal fire spread.
   B. will require the use of more water for fire suppression.
   C. consists of three separate holes cut in a U shape.
   D. will prevent the normal vertical spread of fire.

131. Directing fire streams into ventilation openings can _____ the ventilation process.
   A. defeat
   B. assist
   C. supplement
   D. enhance

132. When exposed to intense heat, a lightweight metal truss can:
   A. maintain its structural integrity.
   B. often contain a fire to a specific area.
   C. be expected to fail in 5 to 10 minutes.
   D. be expected to support firefighting operations for at least 20 minutes.

133. To be **most effective**, trench ventilation operations must be completed before:
   A. fire attack.
   B. the fire reaches the smoldering stage.
   C. the fire reaches the trench.
   D. extinguishment begins.

134. From a life safety point of view, the advantage(s) of proper ventilation for building occupants is/are that it:
   A. improves visibility.
   B. reduces the danger of backdraft explosions.
   C. removes toxic smoke.
   D. All of the above.

135. Ventilation is the:
   A. act of gaining access to secured buildings or areas.
   B. cooling of combustible gases below their flash points.
   C. planned removal of pressure, heat, smoke, and gases through predetermined paths.
   D. removal of smoke and gases from an enclosed area through undetermined paths.

____136. Ventilation should occur:
   A. only if the fire is endangering civilian life safety.
   B. at all structure fires.
   C. at structure fires where the fire is on the top floor.
   D. only at structure fires where the building will be a total loss.

____137. The two types of ventilation are:
   A. natural and mechanical/forced.
   B. hydro and electric.
   C. manual and mechanical.
   D. leeward and windward.

____138. A type of wood framing that has vertical channels going from floor to floor, allowing a fire to travel uninterrupted is a _____ frame.
   A. platform
   B. open
   C. balloon
   D. box

____139. The weight of the building materials and any part of the building permanently attached or built in is the definition of a(n) _____ load.
   A. impact
   B. fire
   C. design
   D. dead

____140. Concrete has excellent _____ strength when it cures.
   A. shear
   B. compressive
   C. torsional
   D. tensile

____141. The search for and extinguishment of hidden fire and placing the building in a safe condition is known as:
   A. overhaul.
   B. secondary search.
   C. size-up.
   D. salvage.

____142. Hidden fires in concealed spaces can often be detected by:
   A. feeling with back of hand.
   B. strategic fan placement.
   C. tearing down the entire wall.
   D. use of salvage techniques.

____143. A fire fighter can often detect hidden fires in a concealed space by:
   A. opening up the entire concealed space.
   B. waiting for flames to appear.
   C. sight, touch, and sound.
   D. smelling for burning material.

144. Firefighters can assist the investigator in all the following ways **except** by:
   A. performing overhaul before the investigator arrives.
   B. reporting unusual fire behavior.
   C. waiting for the investigator to release an area for overhaul.
   D. reporting what witnesses to the fire have said about the fire starting.

145. During overhaul, if fire fighters have a concern about a void space concealing a "hot spot," they should:
   A. check the area periodically without opening the void space.
   B. open void space only if smoke or fire is evident.
   C. open the void space immediately.
   D. check with command for instructions.

146. An observant fire fighter can give the fire investigator all of the following items of information **except**:
   A. the name of the individual(s) who reported the fire.
   B. whether there were signs of a break-in.
   C. how appliances were found.
   D. where the fire was.

147. One way to remove water coming through the ceiling from upper floors is by the use of:
   A. sponges.
   B. chutes.
   C. carryalls.
   D. floor runners.

148. Which of the following is placed on the floor to hold small amounts of water?
   A. Floor runner
   B. Carryall
   C. Catchall
   D. Water chute

149. To form the corners of the basin when constructing a catchall, the fire fighter should lay ends of the side rolls over at a _____ angle.
   A. 30°
   B. 60°
   C. 90°
   D. 180°

150. Methods and operating procedures that reduce fire, water, and smoke damage during and after fires are known as:
   A. overhaul.
   B. size up.
   C. salvage.
   D. a coordinated fire attack.

_____ **151. Directions**: Read the statements below. Then choose the appropriate answer from choices A-D listed below.

  1. Salvage work in the fire service consists of procedures that reduce fire, water, and smoke damage during and after fires.
  2. Overhaul activities consist of all activities that take place after the fire has been extinguished.
  3. Overhaul operations must be completed before salvage.

  A. All statements are true.
  B. Statements 1 and 2 are false; statement 3 is true.
  C. Statement 1 is true; statements 2 and 3 are false.
  D. Statement 1 is false; statements 2 and 3 are true.

_____ **152.** In addition to controlling runoff water, a salvage cover may be used:
  A. to collect debris.
  B. to cover furniture.
  C. as a catchall.
  D. All of the above.

_____ **153.** When making a water chute using a ladder and salvage covers, what other item(s) is/are required?
  A. Halligan tool
  B. Pike poles
  C. Utility rope
  D. Hose line

_____ **154.** The basic premise of salvage operations is:
  A. to prevent fire extension.
  B. to protect fire department property from being damaged at the fire scene.
  C. to separate or protect interior and exterior materials from the harmful environment.
  D. to provide better information to the fire inspector.

_____ **155.** If salvage operations are going on while active suppression operations are taking place, how should the salvage crew be dressed?
  A. Station work uniforms are acceptable.
  B. They should wear full protective clothing, including SCBA.
  C. They should wear full protective clothing less SCBA.
  D. Gloves and a helmet with eyeshield are appropriate attire.

_____ **156.** A carry-all is used for:
  A. removing fire victims.
  B. bringing additional fire equipment to the scene.
  C. catching water that is leaking into an area.
  D. carrying debris out of an area.

___**157.** Which of the following items is a sign that a sheet rock ceiling is becoming saturated with water and may collapse?
   A. The sheet rock will begin to smoke.
   B. The seams will begin to show.
   C. The sheet rock will begin to char.
   D. The sheet rock will start to contract.

___**158.** When the water flow alarm (water gong) sounds, this indicates that:
   A. water has stopped flowing in the system.
   B. heat detection devices have been activated and one may expect the deluge set to begin discharging water momentarily.
   C. water is flowing in the system.
   D. a heat actuating device has been activated and someone should turn the main sprinkler valve to the open position.

___**159.** The flow of water from an individual sprinkler head can be controlled by using:
   A. the main system control valve only.
   B. sprinkler wedges or tongs.
   C. the valve required at each branch line.
   D. individual controls.

___**160.** Sprinkler heads rated for the Ultra High temperature classification are color coded:
   A. yellow.
   B. white.
   C. blue.
   D. orange.

___**161.** A sprinkler head rated for 180 degrees will be color coded:
   A. white.
   B. blue.
   C. red.
   D. green.

___**162.** The temperature rating of a sprinkler head color-coded red is _____ degrees F.
   A. 175 to 225
   B. 250 to 300
   C. 325 to 375
   D. 400 to 475

___**163.** A _____ is commonly used to stop water from flowing from a sprinkler head.
   A. wedge
   B. pipe wrench
   C. halligan
   D. rubber pellet

____164. The space provided for hose on fire apparatus is generally referred to as the:
   A. hose box.
   B. hose load.
   C. hose bed.
   D. hose lay.

____165. Water supply is one of the most critical elements of firefighting because:
   A. of the great expense in obtaining it.
   B. water is the most common extinguishing agent.
   C. water freezes at high temperatures.
   D. of its ability to suffocate a fire.

____166. Available flow/static pressure is the:
   A. rate and quantity of water delivered.
   B. amount of water flowing from the discharge side of the pump.
   C. amount of water required to put out the fire.
   D. amount of water that can be moved to put out the fire.

____167. The <u>most</u> <u>common</u> water distribution system is a _____ system.
   A. pumped
   B. combination pumped/gravity
   C. gravity
   D. tender shuttle

____168. The two major hydrant types are:
   A. wet barrel and dry barrel.
   B. high-pressure and low-pressure.
   C. ground water and surface water.
   D. treated water and untreated water.

____169. Which of the following statements regarding wet barrel and dry barrel hydrants is incorrect?
   A. Dry barrel hydrants are used in areas where freezing temperature could damage the hydrant.
   B. Wet barrel hydrants have water in the barrel up to the valves of each outlet.
   C. Dry barrel hydrants use a valve at the base to control water flow to all outlets.
   D. Wet barrel hydrants are commonly used in the northern parts of the United States (Alaska, Minnesota).

____170. Residual pressure is:
   A. the pressure in the system with no hydrants or water flowing.
   B. the pressure in a system after water has begun flowing.
   C. the level of ground water under the surface.
   D. a device that speeds the unloading of water from a tender.

____171. Dump site selection should be based on:
   A. access to hydrants.
   B. turn-around area for the tenders.
   C. the size of available mains.
   D. the location of the fill site.

_____172. What step **should not** be taken when attempting to lower shuttle time?
   A. Increase efficiency of personnel at fill site.
   B. Increase vehicle speed on highway during shuttle operations.
   C. Increase efficiency of personnel at dump site.
   D. Use jet dumps to increase dumping time.

_____173. A dry chemical extinguisher rated 60-B is capable of extinguishing a _____ flammable liquid pan fire.
   A. 40 ft^2
   B. 60 ft^2
   C. 120 ft^2
   D. 150 ft^2

_____174. A green triangle containing a letter would indicate an extinguisher to be used on _____ fires.
   A. Class A
   B. Class B
   C. Class C
   D. Class D

_____175. The **preferred** method of applying dry chemical agents to flammable liquid spill fires is to:
   A. direct the stream into the flame and allow it to settle.
   B. deflect the stream a minimum of 5 feet in front of the spill to prevent agitation.
   C. direct the stream up-wind and allow it to be blown onto the fire.
   D. direct the stream at the base of the fire using a sweeping motion.

_____176. Extinguishers suitable for Class B fires can be identified by a _____ containing the letter B.
   A. blue circle
   B. green triangle
   C. red square
   D. yellow star

_____177. Extinguishers suitable for Class C fires can be identified by a _____ containing the letter C.
   A. yellow star
   B. green triangle
   C. red square
   D. blue circle

_____178. A multipurpose dry chemical extinguisher is rated for Class _____ fires.
   A. A and B
   B. B and C
   C. A and C
   D. A, B, and C

___179. Fires involving flammable liquids, greases, and gases where the smothering or blanketing effect is needed are _____ fires.
A. Class A
B. Class B
C. Class C
D. Class D

___180. Fires involving combustible metals such as magnesium, titanium, zirconium, sodium, and potassium, are _____ fires.
A. Class A
B. Class B
C. Class C
D. Class D

___181. The manufacture of all _____ extinguishers has been discontinued.
A. CO_2
B. dry chemical
C. inverting
D. pressurized-water

___182. Aqueous film-forming foam extinguishers are suitable for use on _____ fires.
A. Class B and C
B. Class A and B
C. Class D
D. Class A, B, and C

___183. Fire extinguisher classification symbols are displayed by all of the following **except**:
A. color.
B. shape.
C. letter.
D. weight of container.

___184. Portable fire extinguishers are designed to fight unusual fires that are not easily put out by water and are:
A. large room-sized fires.
B. small incipient fires.
C. fires being fed by oxygen.
D. pressurized fuel fires.

___185. The skill of operating a fire extinguisher is fairly easy in _____ basic steps?
A. two
B. seven
C. four
D. twelve

_____**186.** Cold temperatures would have the **greatest** effect on _____ extinguishing agents stored in an extinguisher.
  A. dry powder
  B. water-based agents
  C. dry chemical
  D. carbon dioxide

_____**187.** Pump tank extinguishers are used to apply:
  A. water.
  B. wet chemical.
  C. dry chemical.
  D. carbon dioxide.

_____**188.** Dry chemicals are effective extinguishing agents due to their:
  A. ability to penetrate.
  B. ability to rapidly cool the atmosphere.
  C. coating action.
  D. ability to be applied in extremely windy conditions.

_____**189.** What extinguishing agents are being replaced because they have been banned for destroying the Earth's ozone layer?
  A. Halons or halogenated hydrocarbons
  B. Carbon dioxide
  C. Aqueous Film Forming Foam (AFFF)
  D. Dry chemicals

_____**190.** Firefighters should treat all downed wires as:
  A. energized.
  B. safe if in contact with the ground.
  C. only dangerous if nearby homes have power.
  D. safe if not arcing.

_____**191.** Three important factors in ground cover firefighting are fuel, weather, and:
  A. flames.
  B. topography.
  C. heat.
  D. oxygen

_____**192.** A suppression action taken by a fire fighter around the perimeter of a wildland fire is called:
  A. black line.
  B. direct attack.
  C. cold fire edging
  D. water curtain.

___193. Which of the following statements regarding building inspections is <u>incorrect</u>?
   A. Fire inspectors should not be escorted while performing inspections.
   B. How a building is inspected is not important as long as the method is efficient, systematic, and thorough.
   C. Company officers should insist that the crew be escorted during the inspection.
   D. Under no circumstances should company officers argue with business owners.

___194. During the pre-incident site visit, what information should be obtained and documented?
   A. Built-in fire protection
   B. Access points to the site and interior of the structure
   C. Structure size, height, and number of stories
   D. All of the above are correct.

___195. Fire departments should educate _____ to recognize potential hazards and take appropriate corrective action.
   A. preschoolers
   B. the elderly
   C. adults
   D. citizens of all ages

___196. Defective SCBA cylinder units should be:
   A. repaired by the person who discovers the defect.
   B. removed from service.
   C. put on reserve fire apparatus.
   D. filled to 80% capacity.

___197. Composite SCBA cylinders must be hydrostatically tested every:
   A. year.
   B. three years.
   C. five years.
   D. ten years.

___198. Steel and aluminum cylinders for breathing apparatus should be hydrostatically tested after each _____ -year period.
   A. two
   B. three
   C. four
   D. five

___199. Ropes should be inspected:
   A. by conducting static load and elongation tests.
   B. only when contact with chemicals has occurred.
   C. after each use.
   D. a minimum of every two years.

____**200.** A _____ finish is added to the basic hose load on an apparatus to increase the versatility of the load.
   A. hose load
   B. split hose
   C. dutchman
   D. minuteman

Did you score higher than 80% on Examination I-3? Circle Yes or No <u>in ink</u>.

Feedback Step

Now, what do we do with your Yes and No answers through the Phase I examination process? First, return to any response that has No circled. Go back to the highlighted answers for those examination items missed and read and study the paragraph *preceding* the location of the answer, as well as the paragraph *following* the paragraph where the answer is located. This will expand your knowledge base for the missed question, put it in a broader perspective, and improve associative learning. Remember, we are trying to develop mastery of the required knowledge. Scoring 80% on an examination is good but it **is not** mastery performance. To come out in the top of your group you must score well above 80% on your training, promotion, or certification examination.

Carefully review the Summary of Key Rules for Taking an Examination and Summary of Helpful Hints. Do this review now and at least two additional times prior to taking your next examination.

Helpful Hint

Studying the correct answers for missed items is a critical step in return on effort! The focus of attention is broadened and many times new knowledge is gained by expanding association and contextual learning. During our research and field test, self-study during this step of SAEP resulted in gain scores of 17 points between the first examination administered to the third examination. A gain score of 17 points can move you from the lower middle to the top of the list of persons taking a training, promotion, or certification examination. That is a **competitive edge** and prime example of return on effort in action. Remember: Maximum effort = Maximum results!

Summary of Key Rules for Taking an Examination

<u>Rule 1</u>—Examination preparation is not easy. Preparation is 95% perspiration and 5% inspiration.

<u>Rule 2</u>—Follow the steps very carefully. Do not try to reinvent or shortcut the system. It really works just as it was designed to!

<u>Rule 3</u>— Mark with an "X" any examination items for which you guessed the answer. For maximum return on effort, you should also research any answer that you guessed, even if you guessed correctly. Find the correct answer, highlight it, and then read the entire paragraph that contains the answer. Be honest and mark all questions you guessed. Some examinations have a correction for guessing built into the scoring process. The correction for guessing can reduce your final examination score. If you are guessing, you are not mastering the material.

<u>Rule 4</u>—Read questions twice if you have any misunderstanding and especially if the question contains complex directions or activities.

<u>Rule 5</u>—If you want someone to perform effectively and efficiently on the job, the training and testing program must be aligned to achieve this result.

<u>Rule 6</u>—When preparing examination items for job-specific requirements, the writer must be a subject matter expert with current experience at the level that the technical information is applied.

<u>Rule 7</u>—Good luck = Good preparation.

Summary of Helpful Hints

Helpful Hint—Most of the time your first impression is the best. More than 41% of changed answers during our SAEP field test were changed from a right answer to a wrong answer. Another 33% changed their answer from wrong to wrong. Only 26% of changed answers were from wrong to right. In fact, a number of changed answers resulted in three participants not making a perfect score of 100% when changing one right answer to a wrong one! Think twice before you change your answer. The odds don't appear to be in your favor.

Helpful Hint—Researching correct answers is one of the most important activities in SAEP. Locate the correct answer for all missed examination items. Highlight the correct answer. Then read the entire paragraph containing the answer. This will put the answer in context for you and provide important learning by association.

Helpful Hint—Proceed through all missed examination items using the same technique. Reading the entire paragraph improves retention of the information and helps you develop an association with the material and learn the correct answers. This step may sound simple. A major finding during the development and field testing of SAEP was that you learn from your mistakes.

Helpful Hint—Follow the steps carefully to realize the best return on effort. Would you consider investing your money in a venture without some chance of return on that investment? Examination preparation is no different. You are investing time expecting a significant return for that time. If, indeed, time is money, then you are investing money and are due a return on that investment.

Helpful Hint—Try to determine why you selected the wrong answer. Usually something influenced your selection. Focus on the difference between your wrong answer and the correct answer. Carefully read and study the entire paragraph containing the correct answer. Highlight the answer just as you did for Examination I-1.

Helpful Hint—Studying the correct answers for missed items is a critical step in return on effort! The focus of attention is broadened and many times new knowledge is gained by expanding association and contextual learning. During our research and field test, self study during this step of SAEP resulted in gain scores of 17 points between the first examination administered to the third examination. A gain score of 17 points can move you from the lower middle to the top of the list of persons taking a training, promotion, or certification examination. That is a competitive edge and prime example of return on effort in action. Remember: Maximum effort = Maximum results!

PHASE II

Fire Fighter II

Examination II-1, Surveying Weaknesses

At this point in SAEP you should have the process of self-directed learning using examinations fixed in your mind. Moving through this Phase is accomplished the same as in Phase 1. Do not attempt to skip steps in the process because you understand how SAEP works. This can lead to a weak examination preparation result. The Phase II Fire Fighter exams will be more difficult because of the increased level or required knowledge and skills. You will find that the SAEP methods move you gradually from the simple to the complex.

Do not study prior to taking the examination. The examination is designed to identify your weakest areas in terms of NFPA 1001. There will be steps in the SAEP that require self-study of specific reference materials.

Mark all answers in ink. The reason for this is to make sure no corrections or changes are made. Do not mark through answers or change answers in any way once you have selected the answer. *Doing so indicates uncertainty regarding the answer. Mastery is not compatible with uncertainty.*

Step 1—Take Examination II-1. When you have completed Examination II-1, go to Appendix B and compare your answers with the correct answers. Notice that each answer has reference materials with page numbers. If you missed the correct answer to the examination item, you have a source for conducting your correct answer research.

Step 2—Score Examination II-1. How many examination items did you miss? Write the number of missed examination items in the blank <u>in ink</u>. _____ Enter the number of examination items you guessed in this blank. _____ Go to your personal progress plotter and enter these numbers in the designated locations.

Step 3—Now the learning begins! Carefully research the page cited in the reference material for the correct answer. For instance, if you are using Jones and Bartlett, *Fundamentals of Fire Fighter* Skills, First Edition, go to the page number provided and find the answer.

Following are some of the Rules and Hints repeated from Phase 1.

──────── Rule 3 ────────

Mark with an "X" any examination items for which you guessed the answer. For maximum return on effort, you should also research any answer that you guessed, even if you guessed correctly. Find the correct answer, highlight it, and then read the entire paragraph that contains the answer. Be honest and mark all questions you guessed. Some examinations have a correction for guessing built into the scoring process. The correction for guessing can reduce your final examination score. If you are guessing, you are not mastering the material.

──────── Rule 4 ────────

Read questions twice if you have any misunderstanding, especially if the question contains complex directions or activities.

> **Helpful Hint**
>
> Proceed through all missed examination items using the same technique. Reading the entire paragraph improves retention of the information and helps you develop an association with the material and learn the correct answers. This step may sound simple. A major finding during the development and field testing of SAEP was that you learn from your mistakes.

Examination II-1

Directions

Remove Examination II-1 from the manual. First, take a careful look at the examination. There should be 100 examination items. Notice that a blank line precedes each examination item number. This line is provided for you to enter the answer to the examination item. Write the answer in ink. Remember the rule about changing the answer. Our research has shown that changed answers are often incorrect, and more often than not the answer that is chosen first is correct.

If you guess the answer to a question, place an "X" or a checkmark by your answer. This step is vitally important as you gain and master knowledge. We will explain how we treat the "guessed" items later in SAEP.

Take the examination. Once you complete it, go to Appendix B and score your examination. Once the examination is scored, carefully follow the directions for feedback on the missed and guessed examination items.

_____ 1. When a company officer arrives first on a fire scene, the officer is in command until:
   A. the fire is declared under control.
   B. a chief officer arrives and may choose to assume command.
   C. the chief of the department arrives.
   D. arrival of the senior shift officer.

_____ 2. An employee becomes frustrated because he/she cannot comply with conflicting orders from different bosses. This situation was caused by a violation of:
   A. chain of command.
   B. division of labor.
   C. span of control.
   D. unity of command.

_____ 3. In the fire service, division of labor is necessary to:
   A. assign responsibility.
   B. prevent duplication of effort.
   C. make specific and clear-cut assignments.
   D. All of the above.

_____ 4. Policies are examples of standing plans designed to provide:
   A. staffing requirement guidelines.
   B. guidance for decision making.
   C. problem-solving.
   D. communications.

_____ 5. A procedure is a(n):
   A. guide to thinking.
   B. detailed guide to action.
   C. guide to decision making.
   D. interpretation.

_____ **6.** Which of the following is one of the major functions of the Incident Management System?
   A. PIO
   B. Planning
   C. Liaison
   D. Safety Officer

_____ **7.** The Incident Management System has a number of components that provide the basis for an effective IMS operation. Which of the following is not one of these components?
   A. Common terminology
   B. Unified command
   C. Modular organization
   D. Prefire plans

_____ **8.** Which of the following is not a major functional area of the Incident Management System?
   A. Groups
   B. Planning
   C. Logistics
   D. Operations

_____ **9.** Passengers and drivers/operators of responding emergency vehicles should do all of the following except:
   A. wear hearing protection when noise levels exceed 90 decibels.
   B. ride on the tailboard.
   C. remain seated in their seats with their seat belts fastened.
   D. ride inside a fully enclosed portion of the cab whenever possible.

_____ **10.** Under normal conditions, only a(n) _____ may order multiple alarms or additional resources for large-scale incidents.
   A. logistics officer
   B. planning officer
   C. safety operations officer
   D. incident commander

_____ **11.** As it relates to the Incident Management System (IMS), a division refers to:
   A. fire fighters assigned to a single task.
   B. a part of a strike team.
   C. a geographic location or designation.
   D. a supporting branch for the logistics team.

_____ **12.** Which of the following is not a component of an Incident Management System?
   A. Integrated communications
   B. Predesignated facilities
   C. Modular organization
   D. Independent action plans

____ 13. The arrival report should contain:
  A. a situation evaluation.
  B. the attack mode selected.
  C. the person in command.
  D. All of the above.

____ 14. United States fire loss statistics show that most structure fires, most fire damage, and most injuries and fatalities occur in _____ occupancies.
  A. commercial
  B. industrial
  C. institutional
  D. residential

____ 15. Which of the following items **would not** be found in the risk/benefit philosophy of a risk management plan?
  A. Where no life can be saved, no risk shall be taken by fire fighters.
  B. Situations endangering valued property shall cause fire fighters to take a calculated and weighted risk.
  C. Where no life or valued property can be saved, risk shall be taken by fire fighters.
  D. Significant risk to the life of a fire fighter shall be limited to those situations where the fire fighter can potentially save endangered lives.

____ 16. The type of call, action taken, number of injuries or fatalities, and the property usage information are all entries to be included on an _____ report.
  A. USFA
  B. IMS
  C. ACLS
  D. NFIRS

____ 17. A uniform data collection system used by most departments to track incident information is known as the:
  A. National Fire Incident Reporting System.
  B. National Fire Incident Recording System.
  C. First National Incident Reporting System.
  D. First National Incident Response System.

____ 18. Which of the following statements regarding fire reports is **incorrect**?
  A. Information found in fire reports can be used by insurance companies.
  B. Information reported on fire reports is used by private manufacturing companies.
  C. Fire reports are public records.
  D. NFIRS do not need to be completed for EMS runs.

____ 19. The two **most common** ways the incident commander orders fire fighters to evacuate a structure are to broadcast a radio message and:
  A. page all fire fighters to respond.
  B. implement an accountability system.
  C. contact dispatch to activate PASS device.
  D. sound an audible warning.

_____ **20.** Audible warning devices for emergency evacuation should be:
   A. broadcast several times.
   B. heard for at least 500 feet.
   C. used to announce the need for multiple alarms.
   D. used to give an "all clear" on scene.

_____ **21.** It is important that the communication center be kept advised of the actions taken at emergency scenes. Progress reports should include all of the following <u>except</u>:
   A. change in command location.
   B. exposures present.
   C. direction of fire spread.
   D. number of units in staging.

_____ **22.** Fire departments that operate radio equipment must hold radio licenses from the:
   A. Federal Central Communications.
   B. National Emergency Broadcasting.
   C. Federal Communications Commission.
   D. National Radio Communications.

_____ **23.** A(n) _____ System tracks fire department units on a map as they move around city or town streets.
   A. Low-Jack
   B. Automatic Global Positioning (AGP)
   C. Automatic Vehicle Locating (A.V.L.)
   D. Vehicle Down-Link (V.D.L)

_____ **24.** In fire departments that have access to multiple radio channels, emergency operations should be:
   A. on multichannels also.
   B. run by cell phone so the radio is not tied up.
   C. assigned a separate channel dedicated for use on that scene only.
   D. Both A and C are correct.

_____ **25.** The important difference between Basic 911 and Enhanced 911 is that:
   A. enhanced systems have the capability to provide the caller's telephone number and address.
   B. enhanced systems are used only in rural areas.
   C. basic systems are more reliable than enhanced.
   D. basic systems have the capability to provide the caller's telephone number and address.

_____ **26.** The communications center is:
   A. the ideal place to put unskilled employees.
   B. the hub of the operation.
   C. a low-stress work environment.
   D. the least needed branch of the fire service.

_____ **27.** The report given to the I.C. from an interior crew which tells the I.C. the fire is controlled would be a(n):
   A. size-up report.
   B. progress/status report.
   C. all clear notification.
   D. staging report.

_____ **28.** Which of the following is important to remember in placing a foam line in service using an in-line proportioner?
   A. Check the eductor and nozzle to make sure they are hydraulically compatible (rated for the same flow).
   B. Check to see that the foam concentrate listed on the foam container matches the eductor percentage rating.
   C. Select the proper foam concentrate for the burning fuel involved.
   D. Attach the eductor to a hose capable of efficiently flowing the rated capacity of the eductor and the nozzle.

_____ **29.** Which is the **most common** type of foam proportioner used in the fire service?
   A. Balanced pressure proportioners
   B. In-line eductors
   C. Around-the-pump proportioners
   D. Automatic eductors

_____ **30.** The elements needed to produce quality firefighting foam include:
   A. mechanical aeration, air, water, and concentrate.
   B. air, concentrate, eductor, and CO_2.
   C. proportioner, CO_2, and eductor.
   D. aspiration, subsurface injection, and air.

_____ **31.** Class A foams are essentially wetting agents that _____ of water.
   A. increase the viscosity
   B. increase the resistance
   C. reduce the surface tension
   D. create a higher vaporization point

_____ **32.** Alcohol resistant AFFF (Aqueous Film Forming Foam) can be utilized on hydrocarbon fires at _____ percent proportions.
   A. two
   B. three
   C. ten
   D. nine

_____ **33.** Which one of the following **is not** a method by which foam acts as a suppression agent?
   A. Smothering
   B. Cooling
   C. Inhibiting the chemical chain reaction
   D. Separating

____ 34. Which of the following **is not** a polar solvent?
   A. Alcohol
   B. Acetone
   C. Kerosene
   D. Ketone

____ 35. Petroleum-based fuels are _____ and float on water.
   A. inorganic
   B. hydrocarbons
   C. polar solvents
   D. polymers

____ 36. To produce the proper rate of foam on flammable liquids fire, a(n) _____ is necessary.
   A. high-volume pump
   B. proportioner
   C. fog nozzle
   D. aspirating tip

____ 37. The safest recommended means for a fire fighter to disconnect electrical service to a building is to:
   A. cut the service entrance wire.
   B. pull the meter.
   C. locate the nearest transformer and deactivate it.
   D. shut off the main power breaker/fuse in the panel box.

____ 38. A roof that is elevated in the center and with an angular slope to the edges is called a _____ roof.
   A. butterfly
   B. dome
   C. pitched/gabled
   D. double-angle

____ 39. The **primary** fire hazard in fire resistive construction is the:
   A. structure members.
   B. contents of the structure.
   C. lack of walls.
   D. non-fire resistive construction.

____ 40. Horizontal extension of fire includes all of the following means of heat transfer **except** through:
   A. wall openings by direct flame contact.
   B. open space by radiant heat.
   C. ceilings and floors by direct flame contact.
   D. walls by conduction of heat through pipes.

_____ **41.** The type of construction that has the **greatest** resistance to structural damage by fire is _____ construction.
   A. heavy timber
   B. ordinary
   C. fire resistant
   D. noncombustible

_____ **42.** Spalling of concrete could lead to early collapse in Type I buildings because:
   A. loss of moisture in concrete reduces its fire rating.
   B. the added weight of broken pieces may cause overload.
   C. it could create void spaces.
   D. reinforcing steel is exposed to the heat of the fire.

_____ **43.** What type of construction has structural members (including walls, columns, beams, floors, and roofs) that are made of noncombustible or limited-combustible materials?
   A. Type I
   B. Type III
   C. Type IV
   D. Type V

_____ **44.** Firefighters should know that fire in Type V construction presents:
   A. shortening of steel components.
   B. breakdown of the concrete members due to the heat buildup.
   C. extensive spalling.
   D. high potential for fire extension within the building.

_____ **45.** What type of building construction is made up of solid heavy timber or laminated wood?
   A. Type I
   B. Type II
   C. Type IV
   D. Type V

_____ **46.** Which of the following is a hazard associated with truss and lightweight construction?
   A. If one member fails, the entire truss is likely to fail.
   B. Once a truss fails, the one next to it is likely to fail.
   C. Trusses will begin to fail after five to ten minutes of exposure to fire.
   D. All of the above.

_____ **47.** One of the **most serious** building construction hazards facing fire fighters today is the:
   A. increased use of noncombustible materials.
   B. increased use of lightweight and trussed support systems.
   C. heavy content of fire loading.
   D. presence of combustible furnishings and finishes.

_____ 48. One thing common to all types of trusses is that if one member fails:
A. only that member will fail.
B. the truss next to it will keep it from failing completely.
C. the entire truss is likely to fail.
D. it is unlikely to have a total collapse.

_____ 49. Which of the following is not an indicator of potential building collapse?
A. Prolonged fire operations in fire building
B. Deteriorated mortar between the masonry
C. Walls that appear to be leaning
D. Large amounts of steam coming from ventilation openings

_____ 50. Wire reinforced glass may provide some thermal protection as a separation. However, for the most part, conventional glass:
A. is not an effective barrier to fire extension.
B. is a good barrier to fire extension.
C. will not crack due to the heat.
D. will contain the fire within that area.

_____ 51. A wall used to divide two adjacent structures and also could be used as a fire wall is a _____ wall.
A. partition
B. party
C. veneer
D. cantilever

_____ 52. Directions: Read the following statements, then select your answer from alternatives A-D below.

To ensure that there is little danger of injury, a fire ax should be carried:
1. on the shoulder with the edge pointed toward the ground.
2. with the ax blade away from the body, or protected.
3. with pick-head axes, the pick should be covered with a hand.

A. All three statements are true.
B. Statement 1 and Statement 2 are false; Statement 3 is true.
C. Statement 1 is true and Statements 2 and 3 are false.
D. Statement 1 is false and Statements 2 and 3 are true.

_____ 53. What kind of heat energy is the heat of compression?
A. Chemical
B. Electrical
C. Mechanical
D. Nuclear

____ **54.** Ignition temperature is the **minimum** temperature required to:
  A. cause a fuel to give off vapors in sufficient quantities to form an ignitable mixture with air.
  B. heat a fuel which will produce vapors sufficient to support combustion once ignited.
  C. heat a fuel to begin self-sustained combustion independent of the heating source.
  D. change a liquid to a gas without introducing an outside source of heat.

____ **55.** Pressurized flammable liquids and gases should:
  A. always be extinguished.
  B. not be extinguished unless the fuel can be immediately shut off.
  C. not be extinguished by fire fighters; trained specialists should be called for these fires.
  D. not be extinguished unless the product involved has a vapor density greater than one.

____ **56.** During a fire, in order to achieve the **maximum** efficient use of water when cooling flammable gas storage tanks, fire streams should be directed:
  A. around the tank base.
  B. above the level of the contained liquid.
  C. below the level of the contained liquid.
  D. into the involved tank.

____ **57.** Fires burning at the relief valves or piping should:
  A. be extinguished with water.
  B. be extinguished with foam.
  C. be extinguished with fog streams.
  D. not be extinguished.

____ **58.** When containers of flammable gases are exposed to flame impingement, the water for cooling the container should be applied to cool the:
  A. vapor space.
  B. relief valve.
  C. ends of the tanks.
  D. fire fighters.

____ **59.** Which of the following **is not** one of the normal observations that fire fighters make to assist in determining fire cause?
  A. How the fire reacted to water application
  B. People leaving the fire scene in a hurry
  C. Hindrances to firefighting
  D. The number and location of observers

____ **60.** Firefighters should always remember how they gained entry into a building because:
  A. entry may have an effect on the behavior of the fire.
  B. the fire investigator may want to know how entry was made.
  C. you may need to use the opening in case of a rekindle.
  D. you should always exit the same way you entered.

_____ **61.** Prior to the fire investigator's arrival, fire fighters should _____ any evidence found.
  A. tag and photograph
  B. protect and preserve
  C. collect and package
  D. isolate and remove

_____ **62.** Following a fire of suspicious origin, the fire fighter should:
  A. carefully wash all burned articles with water to clean them off.
  B. remove any possible evidence to the outdoors where it can be properly tagged, identified, and photographed.
  C. have the building owner walk through the building to pick up valuables.
  D. leave suspected evidence where it is found.

_____ **63.** Hidden fire can be checked by using a(n):
  A. detector for different levels of carbon monoxide and oxygen.
  B. Halligan tool to remove the whole wall.
  C. plaster tool from the opposite side of the wall.
  D. electronic/infrared heat sensor.

_____ **64.** Initial recognition and preservation of evidence is the responsibility of the:
  A. fire marshal.
  B. fire inspector.
  C. fire investigator.
  D. fire fighter.

_____ **65.** Once a fire investigator has completed the work required in gathering all of the required evidence and information from a fire scene, a thorough _____ can be done.
  A. ventilation
  B. demobilization
  C. inventory
  D. overhaul

_____ **66.** Which of the following statements regarding fire cause determination is **incorrect**?
  A. Proper overhaul skills will rearrange fire scene evidence.
  B. A fire fighter is an important link in the chain of fire cause determination.
  C. Firefighters may be required to give statements to insurance investigators.
  D. Firefighters should always treat all fire scenes as a crime scene.

_____ **67.** Any vehicles in which a person is entrapped must be _____ to prevent them from shifting and inflicting more damage or injuries.
  A. stabilized
  B. hosed down
  C. removed
  D. left untouched

____ 68. Which of the following tools is essential for stabilizing a vehicle?
   A. Hydraulic jack
   B. Block and tackle
   C. High-pressure air bags
   D. Cribbing

____ 69. If possible, entry to a vehicle for rescue purposes should be made through the:
   A. windshield.
   B. doors.
   C. roof.
   D. trunk/hatch.

____ 70. The phase of vehicle extrication that determines the nature and extent of the overall situation is:
   A. preparation.
   B. disentanglement.
   C. hazard control.
   D. size-up.

____ 71. Of the situations listed below, which **would not** apply to a shoring situation?
   A. Vehicle stabilization
   B. Earth openings
   C. Cave-ins
   D. Building collapse

____ 72. Powered hydraulic tools open and close by use of:
   A. air.
   B. fluid.
   C. steam.
   D. mechanical advantage.

____ 73. When using pneumatic air bags to perform a lift, one should never stack **more** than _____ bags.
   A. two
   B. three
   C. four
   D. five

____ 74. Which of the following **would not** be considered a confined space?
   A. Utility vault
   B. Septic tank
   C. Trench box
   D. Storage tank

____ 75. The process of erecting materials such as wood panels, timber, or jacks to strengthen a wall or prevent further collapse is known as:
   A. shoring.
   B. cribbing.
   C. packing.
   D. supporting.

_____ 76. The <u>most</u> <u>common</u> hazard encountered at a confined space incident is:
  A. an oxygen deficient atmosphere.
  B. engulfment.
  C. collapse of structure.
  D. entrapment.

_____ 77. An example of a manually activated fire alarm component is a:
  A. thermally sensitive device.
  B. flame detector.
  C. water flow detector.
  D. pull station.

_____ 78. A special communications device which allows the hearing or speech impaired to communicate via telephone is known as a _____ system.
  A. commercial phone
  B. TDD/TTY text phone
  C. direct line
  D. wireless

_____ 79. When the water source <u>does</u> <u>not</u> have adequate elevation to create proper pressure for gravity flow, it is necessary to use:
  A. larger size pipes in the mains.
  B. pumps to raise the system's pressure.
  C. computer-controlled pressure regulators.
  D. negative coefficient of friction loss.

_____ 80. Manual detection systems typically have two problems. The first is that many are local only, and the second is that:
  A. they are very technical to use.
  B. they are placed in closets to reduce false alarms.
  C. building occupants need a key to operate system.
  D. a person must be present, awake, and alert to discover the fire and activate the system.

_____ 81. Heat detectors:
  A. cannot be used as part of a suppression system.
  B. are slow to activate.
  C. are expensive to install and operate.
  D. are responsible for most false alarms.

_____ 82. Smoke detectors work primarily on the principles of photoelectricity and:
  A. rate of rise.
  B. fixed temperature.
  C. ionization.
  D. laser beam.

_____ 83. The purpose of the fire department connection to a sprinkler system is to:
   A. supplement the water supply while maintaining operational pressure.
   B. provide water, since most systems are dependent on the fire department for water supply.
   C. boost the water to upper stories, since most water pressure is not sufficient to supply water above the sixth floor.
   D. add water pressure to the system because normal water distribution is inadequate when less than three heads are activated.

_____ 84. _____ sprinkler systems should be used in buildings where piping may be subjected to freezing temperatures.
   A. Wet-pipe
   B. Dry-pipe
   C. Deluge
   D. Residential

_____ 85. The common classifications of sprinkler systems include all of the following except _____ systems.
   A. preaction
   B. residential
   C. deluge
   D. external-supply

_____ 86. The full name of a PIV is _____ valve.
   A. position-indicator
   B. post-indicator
   C. plant-indicator
   D. positive-indicator

_____ 87. The _____ system is equipped with all sprinkler heads of the open type.
   A. wet-pipe
   B. dry-pipe
   C. deluge
   D. preaction

_____ 88. A fire department connection to a sprinkler system enables fire fighters to:
   A. connect handlines for attacking the fire.
   B. drain water from the system.
   C. increase volume and pressure.
   D. test the system.

_____ 89. The full name of an OS&Y valve is _____ valve.
   A. open stem and yoke
   B. outside, shut, and yoke
   C. outside shield and yoke
   D. outside screw and yoke

_____ 90. Every sprinkler system should be equipped with a main control valve located between the:
   A. riser and the branches.
   B. source of water supply and the sprinkler system.
   C. cross mains and the riser.
   D. fire department connection and the riser.

_____ 91. The control valve for a sprinkler system may be located at the system or outside the building. This valve should always be a(n) _____ valve.
   A. check
   B. indicating-type
   C. inspector's test
   D. quarter-turn

_____ 92. Which of the following **should not** be used as a control valve for an automatic sprinkler system?
   A. Post indicator valve
   B. Wall post indicator valve
   C. Outside screw and yoke
   D. Gate valve

_____ 93. An OS&Y valve is open when the threads are:
   A. retracted.
   B. extended.
   C. reversed.
   D. crossed.

_____ 94. A riser is a _____ pipe that supplies a sprinkler system.
   A. lateral
   B. horizontal
   C. diagonal
   D. vertical

_____ 95. A sprinkler deflector converts a water stream into:
   A. spray.
   B. steam.
   C. a fine mist.
   D. a solid stream.

_____ 96. A network of intermediate-sized pipe that reinforces the overall grid system by forming loops that interlock primary feeders **best** defines:
   A. primary loop.
   B. secondary feeders.
   C. distributors.
   D. grid network.

_____ 97. Which of the following pressures can be measured by a pitot tube?
   A. Static
   B. Normal operating
   C. Residual
   D. Flow

_____ **98.** A flow test from a fire hydrant will reveal:
   A. a hydrant's coefficient of discharge.
   B. how much flow pressure is available.
   C. how much water will flow from any hydrant in the grid at any given time.
   D. the size of the main on which that particular hydrant is installed.

_____ **99.** <u>Directions</u>: Read the statements below. Select your answer from alternatives A-D below.

   1. Flow pressure is only measured through a nozzle.
   2. Static pressure is stored potential energy available to move water through pipes, hoses, fittings, and adapters.
   3. Residual pressure is that part of the total available pressure that <u>is</u> <u>not</u> used to overcome friction or gravity while forcing water through pipe, fittings, fire hose, and adapters.

   A. Statements 1, 2, and 3 are true.
   B. Statement 1 is true; statements 2 and 3 are false.
   C. Statements 1 and 3 are false; statement 2 is true.
   D. Statement 1 is false; statements 2 and 3 are true.

_____ **100.** When measuring flow pressure with a hand-held pitot gauge, the tube should be held:
   A. even with the end of the hydrant outlet.
   B. at a distance equal to one-half the diameter of the outlet.
   C. at a distance equal to the diameter of the outlet.
   D. approximately two inches from the end of the outlet.

Did you score higher than 80% on Examination II-1? Circle Yes or No <u>in</u> <u>ink</u>.
(We will return to your Yes or No answer to this question later in SAEP.)

Examination II-2, Adding Difficulty and Depth

During Examination II-2, progress will be made in developing depth of knowledge and skills. Reminder: Follow the steps carefully to realize the best return on effort.

Step 1—Take Examination II-2. When you have completed Examination II-2, go to Appendix B and compare your answers with the correct answers.

Step 2—Score Examination II-2. How many examination items did you miss? Write the number of missed examination items in the blank <u>in ink</u>. _____ Enter the number of examination items you guessed in this blank. _____ Go to your personal progress plotter and enter these numbers in the designated locations.

Step 3—Once again the learning begins. During the feedback step use Appendix B information for Examination II-2 to research the correct answer for items you missed or guessed. Highlight the correct answer during your research of the reference materials. Read the entire paragraph that contains the correct answer.

Examination II-2

Directions

Remove Examination II-2 from the manual. First, take a careful look at the examination. There should be 100 examination items. Notice that a blank line precedes each examination item number. This line is provided for you to enter the answer to the examination item. Write the answer **in ink**. Remember the rule about changing the answer. Our research has shown that changed answers are often incorrect, and, more often than not, the answer that is chosen first is correct.

If you guess the answer to a question, place an "X" or a checkmark by your answer. This step is vitally important as you gain and master knowledge. We will explain how we treat the "guessed" items later in SAEP.

Take the examination. Once you complete it, go to Appendix B and score your examination. Once the examination is scored, carefully follow the directions for feedback on the missed and guessed examination items.

_____ 1. The Incident Management System should:
   A. be fully implemented for all situations.
   B. be initiated by the first fire unit on the scene.
   C. provide procedures that perfectly fit all departments.
   D. eliminate the need for mutual aid assistance.

_____ 2. The tracking of personnel working at an incident requires a system that is standardized for every incident to establish:
   A. accountability.
   B. chain of command.
   C. unity of command.
   D. span of control.

_____ 3. In the Incident Management System, the _____ has authority over the ordering, releasing, and controlling resources.
   A. Incident Commander
   B. Logistics Officer
   C. Operations Officer
   D. Staging Officer

_____ 4. Within the Incident Management System, the positions of Safety, Liaison, and Information are:
   A. divisions.
   B. groups.
   C. functional areas.
   D. command staff positions.

_____ 5. In order for the Incident Management System to function properly, it must contain all of the following components **except**:
   A. common terminology.
   B. integrated communications.
   C. all personnel from a single agency.
   D. consolidated incident action plans.

_____ 6. Under the Incident Management System, the _____ is responsible for implementing the tactical assignments to meet the strategic goal.
A. Incident Commander
B. Operations Chief
C. Planning Chief
D. Safety Officer

_____ 7. Under the Incident Management System, the _____ Officer is responsible for providing factual and accurate information to the media.
A. Safety
B. Liaison
C. Staffing
D. Public Information

_____ 8. Which of the following statements regarding fire incident reports is incorrect?
A. All states in the U.S. participate in the NFIRS program.
B. They are a legal document even if they are not signed.
C. The address location must include both owner and occupant.
D. Local fire information becomes part of the national database.

_____ 9. Fire incident information is reported to the:
A. U. S. Fire Administration.
B. State Fire Marshal's office.
C. insurance companies, if requested.
D. All of the above.

_____ 10. Status reports should be made ten minutes into the incident and at _____ minute intervals thereafter until the incident is under control.
A. 5 - 10
B. 20 - 30
C. 10 - 15
D. exactly 30

_____ 11. The initial report given by first arriving companies is called a:
A. progress report.
B. status report.
C. size-up.
D. personnel arrival report.

_____ 12. For application of aqueous film-forming foam, eductors or proportioners operate on a(n) _____ principle.
A. Bernoulli's
B. Venturi
C. induction
D. pressure

_____ 13. Foams in use currently are of the mechanical type and must be _____ and _____ before they can be used.
A. proportioned, blended
B. stirred, aerated
C. mixed, proportioned
D. proportioned, aerated

_____ 14. Production of an adequate amount of bubbles to form an effective foam blanket is the definition of:
A. proportioning
B. aeration
C. mixing
D. blending

_____ 15. To ensure maximum effectiveness, use foam concentrates _____ the specific percentage for which they are intended to be proportioned.
A. within 10% greater than
B. only at
C. plus or minus 5% of
D. within plus or minus 2% of

_____ 16. Firefighting foam solution is _____ percent water.
A. 95 to 98.6
B. 80 to 85.5
C. 94 to 99.9
D. 80 to 90.7

_____ 17. The **most effective** type of foam for use on polar solvents is:
A. alcohol-resistant.
B. Class A foam.
C. low/high expansion foam.
D. FFFP.

_____ 18. Foams designed for Class B fuel fires are used at _____ percent concentration.
A. 6 to 10
B. 2 to 8
C. 5 to 12
D. 1 to 6

_____ 19. At flammable liquid spills, fire apparatus should be positioned:
A. downhill and downwind.
B. downhill and upwind.
C. uphill and downwind.
D. uphill and upwind.

_____ 20. Which of the following **is not** a hydrocarbon?
A. Ketone
B. Kerosene
C. Gasoline
D. Fuel/heating oil

____ **21.** At 212°F, water expands to _____ times its original volume when it converts to steam.
   A. 500
   B. 970
   C. 1700
   D. 9000

____ **22.** Natural roof ventilation openings consist of:
   A. cutting a 4 ft. X 4 ft. hole.
   B. trench cutting and smoke ejectors.
   C. scuttle hatches, skylights, and stairwell openings.
   D. fire streams from aerial ladders directed across a ventilation opening.

____ **23.** If a ventilation opening is made directly above a fire, it will tend to _____ the fire.
   A. spread
   B. mushroom
   C. localize
   D. extinguish

____ **24.** Which of the following is considered a natural or normal roof opening?
   A. Parapet
   B. Skylight
   C. Soffit
   D. Fire stop

____ **25.** Whenever possible, forced ventilation should be directed:
   A. on the upward side.
   B. on the windward side.
   C. in the same direction as the wind.
   D. in the same direction as master stream operations.

____ **26.** Before cutting an opening in a roof, fire fighters should:
   A. inspect their cutting tools for sharpness.
   B. check for natural or existing openings.
   C. open all top windows on the windward side of the building.
   D. open all bottom windows on the leeward side of the building.

____ **27.** For purposes of ventilation, a large opening at least _____ feet is much better than several small ones.
   A. 1 X 1
   B. 2 X 2
   C. 3 X 3
   D. 4 X 4

____ **28.** Smoke may be removed from a burning building by controlling heat currents. This type of ventilation is called _____ ventilation.
   A. water-fog
   B. negative pressure
   C. positive pressure
   D. All of the above.

___ 29. Before beginning overhaul, it is vital to make sure the building is:
   A. completely saturated with water.
   B. structurally safe.
   C. free of toxic gases and smoke.
   D. thoroughly dewatered.

___ 30. If there is adequate ventilation, steam from a water fog will _____ the toxic gases present.
   A. compress
   B. absorb
   C. displace
   D. contain

___ 31. Solid streams are preferred whenever:
   A. a large volume of smoke is present.
   B. reach and penetration are needed.
   C. fire fighters need a protective curtain.
   D. forced ventilation is necessary.

___ 32. Knowing the effect of fire on common building materials is important since it provides fire fighters with _____ during fire fighting operations at a particular occupancy.
   A. detailed actions to perform
   B. an idea of what to expect
   C. policies for future direction
   D. exact job tasks to perform

___ 33. Tenders combined with _____ can efficiently provide large volumes of water to a fire ground operation.
   A. large-diameter hose
   B. automatic nozzles
   C. portable water tanks
   D. ladder trucks

___ 34. What device speeds the process of dumping a load of water from a tender?
   A. Jet pump
   B. Pitot gauge
   C. Dump/valve
   D. Venturi pump

___ 35. A forward staging area for high-rise fires is usually established _____ floors below the fire floor.
   A. 1
   B. 2
   C. 3
   D. 4

___ 36. A type of wood framing that has vertical channels going from floor to floor, allowing a fire to travel uninterrupted is a _____ frame construction.
   A. platform
   B. open
   C. balloon
   D. box

___ 37. As fire fighters approach a structure that is going to be searched, they should consider the time of day and:
   A. familiarize themselves with the type of building construction.
   B. anticipate occupancy (residential or commercial).
   C. location of doors and windows for emergency exit.
   D. All of the above.

___ 38. When performing roof ventilation, the **primary** hole should be cut:
   A. directly over the fire, if possible.
   B. as far away from the fire as possible.
   C. directly in the fire fighters' escape path.
   D. as close to the edge of the roof as practical.

___ 39. Which of the following factors must be taken into consideration when fighting a structure fire?
   A. Building construction
   B. Length of time the fire has been burning
   C. Occupancy type
   D. All of the above.

___ 40. With respect to tactical considerations, the acronym RECEO stands for:
   A. Rapid, Exit, Company, Emergency, and Orders.
   B. Rescue, Exposures, Confinement, Extinguishment, and Overhaul.
   C. Real, Emergencies, Can, Extend, and Operations.
   D. Rescue, Extinguishment, Confinement, Exposures, and Overhaul.

___ 41. At a **minimum**, fire fighters must work in teams of _____ when entering an involved structure?
   A. two
   B. three
   C. seven
   D. five

___ 42. Which of the following statements is **incorrect**?
   A. Basement fires are more hazardous due to fuel cells being located in them.
   B. Basement fires are more punishing on fire crews.
   C. Ventilation holes in basement fires should be located over the fire.
   D. Cellar nozzles can be used through windows in basement fires.

____ 43. Which of the following statements is <u>incorrect</u>?
   A. The secondary search is the most dangerous.
   B. Searching a building is completed in two different operations-primary and secondary search.
   C. During primary search, the team is often ahead of attack lines and may be above the fire.
   D. The primary search takes place in a rapid but thorough manner in areas most likely to have victims.

____ 44. Successful fire attack on structures should be:
   A. coordinated with other activities, such as ventilation and rescue.
   B. only attempted by certified fire fighters.
   C. coordinated with overhaul operations to protect contents on the fire floor.
   D. All of the above.

____ 45. Which of the following statements regarding trench cut is <u>incorrect</u>?
   A. The trench cut is an offensive action.
   B. The trench cut is 2-4 feet wide.
   C. It is not opened until the entire cut is complete.
   D. It is made in coordination with interior crews.

____ 46. LPG is _____ than air.
   A. 1.5 times lighter
   B. 2 times lighter
   C. 1.5 times heavier
   D. .5 times heavier

____ 47. During overhaul, fire fighters should wear:
   A. boots and gloves; coats and helmets are unnecessary.
   B. lightweight clothing, due to residual heat.
   C. full protective gear, including SCBA.
   D. full protective gear; SCBA is not needed.

____ 48. When coordinating an interior fire attack on a Class A fire, which of the following <u>would not</u> be considered?
   A. Water supply for back up line
   B. Ventilation
   C. Rapid Intervention Team location
   D. Rescue problems

____ 49. A BLEVE:
   A. most commonly occurs when flames contact the relief valve.
   B. can occur when insufficient water is applied to keep the tank cool.
   C. is a slow deterioration of the tank.
   D. is a condition caused by consolidation of vaporization.

____ 50. An increase in the intensity of sound or fire issuing from a relief valve may indicate:
   A. the relief valve is clogged.
   B. rupture of the vessel is imminent.
   C. the fire is burning out.
   D. the tank is cooling down.

____ 51. Proper protective clothing for a fire attack crew on a LPG cylinder includes all of the following except:
   A. SCBA.
   B. thermal protective clothing.
   C. Level A protective clothing.
   D. protective hoods.

____ 52. Clues that help determine where a fire started include all of the following except:
   A. the depth of char on material.
   B. "V" pattern.
   C. area of heaviest damage.
   D. area of heaviest water damage.

____ 53. A "V" pattern is best described as:
   A. an observation cut in the roof of a building.
   B. the shape a stream takes when leaving a straight tip nozzle.
   C. a design that smoke and fire makes on the interior and exterior walls of a fire building.
   D. a cut used in forcible entry to open a rolling door.

____ 54. Measuring the depth of char is performed on burned:
   A. plastic.
   B. steel.
   C. aluminum.
   D. wood.

____ 55. Which of the following would not be a "type" or form of evidence that could be found at a fire scene?
   A. Rolled rags found on the floor leading to the point of origin
   B. Pictures missing from the fire room
   C. Sprinkler system OS&Y stem in the "out" position
   D. Unusual odor in the house.

____ 56. Which of the following devices is designed to provide the maximum in spinal immobilization?
   A. Stretcher
   B. Blanket
   C. Backboard
   D. Litter

57. _____ glass is hardened glass designed to shatter into small pieces.
   A. Tempered
   B. Plastic-coated
   C. Laminated
   D. Shock-resistant

58. A passenger vehicle has several types of glass, including:
   A. plate, safety, and laminated.
   B. tempered and laminated.
   C. safety and laminated.
   D. shock-resistant.

59. The key to an efficient extrication operation is proper _____ of the situation.
   A. dispatch
   B. assessment
   C. triage
   D. assignment

60. If rescuers are able to get into a vehicle, their first action should be to:
   A. begin extraction.
   B. place a backboard.
   C. assess/protect the victim(s).
   D. remove broken glass.

61. You are operating as a member of a rescue company at the scene of an auto extrication. The scene, vehicle, and the patient(s) are stabilized, all hazards are controlled, and access and disentanglement have been accomplished. The next step in the extrication process should be:
   A. termination and post-incident analysis.
   B. patient packaging and patient removal.
   C. transporting patient(s) to appropriate facilities.
   D. initiating extrication operations on another vehicle, if needed.

62. Which of the following statements regarding air bags is <u>incorrect</u>?
   A. Bags must be on or against a solid base.
   B. You must crib as you lift.
   C. Do not stack more than two bags.
   D. When stacking two bags of different sizes the larger bag goes on top.

63. When extricating a patient from a motor vehicle crash, displacement/disentanglement refers to the:
   A. relocating of major parts (i.e., doors, roof, dash).
   B. cutting off of components (i.e., brake pedal, steering wheel).
   C. bending of sheet metal or components.
   D. actual taking apart of the vehicle components.

64. When lifting rescued persons or heavy objects fire fighters should:
   A. use their back to lift, not their legs.
   B. use their legs to lift not, their back.
   C. try to twist and reach at the same time.
   D. squat and lift with shoulders and arms.

_____ 65. When using the Blanket Drag, patients should always be dragged:
   A. feet first.
   B. head first.
   C. sideways.
   D. Either B or C is correct

_____ 66. What is the <u>first</u> choice of action when involved in an ice rescue?
   A. Throwing a line or reaching with a pole
   B. Pushing a long ladder over the ice
   C. Wearing a PFD to walk onto the ice
   D. Paddling a light boat out to the victim

_____ 67. Rapid Intervention Crew is defined as:
   A. any combination of single resources assembled for an assignment.
   B. the designation for a set number of resources of the same type and kind.
   C. a company designated to search for and rescue trapped fire fighters.
   D. a designated group that is used for rapid knock down of wildland fires.

_____ 68. Which of the following sprinkler system components is used to limit water flow to one direction?
   A. OS&Y valve
   B. Check valve
   C. Control valve
   D. Butterfly valve

_____ 69. _____ systems remain the most reliable of all fire protection devices.
   A. Foam
   B. Dry chemical
   C. Automatic sprinkler
   D. Standpipe

_____ 70. <u>Directions</u>: Read the statements below. Select your answer from alternatives A-D below.

   1. The typical rating for sprinkler head in light hazard occupancy is 165°F
   2. It is acceptable to use a pendant-type sprinkler head in the upright position.
   3. Sprinkler heads are color coded by their temperature ratings.

   A. All statements are true.
   B. Statement 1 is true; statements 2 and 3 are false.
   C. Statement 1 is false; statements 2 and 3 are true.
   D. Statement 2 is false; statements 1 and 3 are true.

_____ 71. Under normal circumstances, the air pressure gauge on a dry-pipe sprinkler system will read _____ the water pressure gauge.
   A. the same as
   B. higher than
   C. lower than
   D. almost double

72. A dry-pipe sprinkler system is equipped with two different types of gauges. The air-pressure gauge is located _____ the clapper valve and the water-pressure gauge is located _____ the clapper.
   A. to the left side; to the right side of
   B. above; below
   C. below; above
   D. to the right side; to the left side of

73. A floor-plan sketch or diagram of a structure shows:
   A. a detailed view of the construction features of each floor and fire protection features.
   B. an outline of each floor of the building, including walls, partitions, openings, fire protection features, and areas of life hazard.
   C. a cutaway view of a particular portion of a building in relation to streets and hydrants.
   D. the building and grounds as they are actually seen in depth by the eye.

74. A sketch depicting the general arrangement of a property in reference to streets, adjacent properties, and other important features is known as a:
   A. plot/site plan.
   B. sectional view.
   C. floor plan.
   D. blueprint.

75. In single family residences, _____ are an important part of an early warning system, in order to increase the chance of survival.
   A. signaling devices
   B. manual pull stations
   C. smoke detectors
   D. supervisory devices

76. Before conducting a pre-incident survey and inspection for the purpose of prefire planning, the occupant should be notified in advance so:
   A. the occupant can correct any fire hazards before the fire department arrives.
   B. there will be minimum inconvenience to the owner, occupant, and fire department.
   C. the owner can have fire extinguishers filled.
   D. someone will be at the location to fill out fire department forms.

77. Direct pumping water systems are those in which water:
   A. moves directly into the distribution system by gravity flow.
   B. is supplied directly into the distribution system from elevated storage tanks.
   C. is pumped directly into the distribution system with no elevated storage.
   D. is pumped directly through the distribution system back into the main water supply.

_____ 78. In a water supply system, the size of the water mains from the largest to the smallest are:
   A. primary, distributor, secondary.
   B. distributor, secondary, primary.
   C. secondary, primary, distributor.
   D. primary, secondary, distributor.

_____ 79. Residual pressure is defined as:
   A. stored potential energy available.
   B. that part of the total available pressure that is not used to overcome friction or gravity.
   C. forward velocity pressure at the point of discharge.
   D. the minimum pressure required in a residential area.

_____ 80. The forward velocity pressure at a discharge opening that is recorded by a pitot tube and gauge while water is flowing is known as _____ pressure.
   A. flow
   B. static
   C. normal operating
   D. residual

_____ 81. The usual pressure on a water distribution system during periods of ordinary consumption demand is known as _____ pressure.
   A. normal operating
   B. residual
   C. atmospheric
   D. static

_____ 82. The part of the total pressure that is lost when water is forced through pipe, fittings, fire hose, and adapters is called:
   A. residual pressure.
   B. flow pressure.
   C. friction loss.
   D. static pressure.

_____ 83. The following figure depicts a _____ hydrant.
   A. dry-barrel
   B. wet-barrel
   C. drafting
   D. dry

____ **84.** A rate-of-rise detector will respond to a rapid change in temperature. This type of detector is:
  A. manually activated.
  B. a type of auxiliary protective signal.
  C. faster to react than a fixed temperature detector.
  D. located by the annunciator panel.

____ **85.** How does a deluge system differ from other types of sprinkler systems?
  A. It has air in the system until activated.
  B. It is installed in floors not ceilings.
  C. All sprinkler heads are open without any fusible heads.
  D. All of the above.

____ **86.** Which statement regarding residential sprinklers is **incorrect**?
  A. They are smaller but more expensive versions of wet- or dry-pipe sprinkler systems.
  B. The water supply is combined with the domestic system.
  C. The use of plastic pipe is allowed.
  D. They are designed to control the level of fire involvement such that residents can escape.

____ **87.** Masonry load-bearing walls are **most commonly** found in _____ construction.
  A. ordinary
  B. frame
  C. fire resistive
  D. noncombustible

____ **88.** A connecting plate used in truss construction that can be made of flat steel stock, light gauge metal, or plywood is the definition of a:
  A. joint.
  B. gusset plate.
  C. column.
  D. joist.

____ **89.** During a fire, the control valve on a sprinkler system should be closed:
  A. as soon as the fire department arrives.
  B. after ensuring the fire is out or completely under control.
  C. prior to advancing hoselines into fire area.
  D. when building occupants decide fire is out.

____ **90.** A post indicator valve (PIV) is:
  A. a device to speed the operation of the dry pipe valve by detecting the decrease in air pressure.
  B. designed to control the head pressure at the outlet of a standpipe system to prevent excessive nozzle pressures in hoselines.
  C. a control valve that is mounted on a post case with a small window reading either "open" or "shut."
  D. a control valve that is mounted on a wall in a metal case with a small window reading either "open" or "shut."

_____ 91. Portable power plants should be run at least once a week while:
  A. watching for incoming amp drops.
  B. inspecting connected electrical cords for frays and/or exposed wires.
  C. inspecting the spark plug and plug wire.
  D. powering an electrical device.

_____ 92. Electrical _____ are used when multiple connections are needed.
  A. cords
  B. junction boxes
  C. inverters
  D. power take offs (PTOs)

_____ 93. Overtaxing the power unit will do all of the following **except**:
  A. give poor lighting.
  B. damage the power unit.
  C. increase power to other tools.
  D. restrict the operation of other electrical tools.

_____ 94. Lighting equipment can be divided into two categories:
  A. inverters and generators.
  B. emergency and nonemergency.
  C. auxiliary and installed.
  D. fixed and portable.

_____ 95. A record should be kept on each section of fire hose; information consisting of _____ should be recorded.
  A. the gallons per minute flowed during the test
  B. the annual number of fire responses at which it was used
  C. the date and results of the annual test
  D. number of threads per inch on each coupling

_____ 96. Of the following, the information that **would not** be recorded on a hose log is:
  A. the date of the test.
  B. the person conducting the test.
  C. failure or pass.
  D. the source of water used during testing

_____ 97. To prevent a pressurized hose from injuring a fire fighter, a gate valve should be placed in line with a _____ hole driller in the gate.
  A. ⅛ inch
  B. ¼ inch
  C. ½ inch
  D. ¾ inch

_____ **98.** The following figure depicts a _____ hydrant.
   A. dry-barrel
   B. wet-barrel
   C. drafting
   D. dry

_____ **99.** The term for taking water from a static source using fire department pumpers is:
   A. feeding.
   B. drafting.
   C. distributing.
   D. flowing.

_____ **100.** Failure to open a dry-barrel hydrant fully will result in a reduced amount of available water and will contribute to:
   A. sedimentation.
   B. susceptibility to freezing.
   C. difficulty in draining the main.
   D. ground erosion.

Did you score higher than 80% on Examination II-2? Circle Yes or No <u>in ink</u>.
(We will return to your Yes or No answer to this question later in SAEP.)

Examination II-3, Surveying Weaknesses and Improving Examination-Taking Skills

This examination section is designed to identify your remaining weaknesses in areas of the NFPA 1001. Examination II-3 is randomly generated and contains examination items you have taken before as well as new ones. There will be steps in the SAEP that require self-study of specific reference materials.

Mark all answers in ink. The reason for this is to make sure no corrections or changes are made. Do not mark through answers or change answers in any way once you have selected your answer.

Step 1—Take Examination II-3. When you have completed Examination II-3, go to Appendix B and compare your answers with the correct answers. Notice that reference materials with page numbers are provided for each answer. If you missed the examination item, you have a source for researching the correct answer.

Step 2—Score Examination II-3. How many examination items did you miss? Write the number of missed examination items in the blank <u>in ink</u>. _____ Enter the number of examination items you guessed in this blank. _____ Go to your personal progress plotter and enter these numbers in the designated locations.

Step 3—Now you will begin reinforcing what you have learned! During the feedback step, research the correct answer using Appendix B information for Examination II-3. Highlight the correct answer during your research of the reference materials. Read the entire paragraph containing the correct answer.

Examination II-3

Directions

Remove Examination II-3 from the manual. First, take a careful look at the examination. There are 150 examination items. Notice that a blank line precedes each examination item number. This line is provided for you to enter your answer to the examination item. Write the answer **in ink**. Remember the rule about changing the answer. Our research has shown that changed answers are most often incorrect, and, more often than not, the one that is chosen first is most often correct.

If you guess the answer to a question, place an "X" or checkmark by your answer. This step is vitally important for gaining and mastering knowledge. We will explain how we treat the "guessed" items later in SAEP.

Take the examination. Once you complete it, go to Appendix B and score your examination. Once the examination is scored, carefully follow the directions for feedback on the missed and guessed examination items.

_____ 1. When a company officer arrives first on a fire scene, the officer is in command until:
   A. the fire is declared under control.
   B. a chief officer arrives and may choose to assume command.
   C. the chief of the department arrives.
   D. arrival of the senior shift officer.

_____ 2. An employee becomes frustrated because he/she cannot comply with conflicting orders from different bosses. This situation was caused by a violation of:
   A. chain of command.
   B. division of labor.
   C. span of control.
   D. unity of command.

_____ 3. Which of the following is one of the major functions of the Incident Management System?
   A. PIO
   B. Planning
   C. Liaison
   D. Safety Officer

_____ 4. The Incident Management System has a number of components that provide the basis for an effective IMS operation. Which of the following **is not** one of these components?
   A. Common terminology
   B. Unified command
   C. Modular organization
   D. Prefire plans

_____ 5. As it relates to the Incident Management System (IMS), a division refers to:
A. fire fighters assigned to a single task.
B. a part of a strike team.
C. a geographic location or designation.
D. a supporting branch for the logistics team.

_____ 6. Within the Incident Management System, the positions of Safety, Liaison, and Information are:
A. divisions.
B. groups.
C. functional areas.
D. command staff positions.

_____ 7. Under the Incident Management System, the _____ Officer is responsible for providing factual and accurate information to the media.
A. Safety
B. Liaison
C. Staffing
D. Public Information

_____ 8. The arrival report should contain:
A. a situation evaluation.
B. the attack mode selected.
C. the person in command.
D. All of the above.

_____ 9. The type of call, action taken, number of injuries or fatalities, and the property usage information are all entries to be included on an _____ report.
A. USFA
B. IMS
C. ACLS
D. NFIRS

_____ 10. A uniform data collection system used by most departments to track incident information is known as the:
A. National Fire Incident Reporting System.
B. National Fire Incident Recording System.
C. First National Incident Reporting System.
D. First National Incident Response System.

_____ 11. Fire departments that operate radio equipment must hold radio licenses from the:
A. Federal Central Communications.
B. National Emergency Broadcasting.
C. Federal Communications Commission.
D. National Radio Communications.

____ **12.** The _____ System can significantly shorten response time or enable a dispatcher to handle a greater volume of calls.
   A. Voice Recording
   B. Radio Logging
   C. Wireless Fax
   D. Computer-Aided Dispatch

____ **13.** In fire departments that have access to multiple radio channels, emergency operations should be:
   A. on multichannels also.
   B. run by cell phone so the radio is not tied up.
   C. assigned a separate channel dedicated for use on that scene only.
   D. Both A and C are correct.

____ **14.** The important difference between Basic 911 and Enhanced 911 is that:
   A. enhanced systems have the capability to provide the caller's telephone number and address.
   B. enhanced systems are used only in rural areas.
   C. basic systems are more reliable than enhanced.
   D. basic systems have the capability to provide the caller's telephone number and address.

____ **15.** The communications center is:
   A. the ideal place to put unskilled employees.
   B. the hub of the operation.
   C. a low-stress work environment.
   D. the least needed branch of the fire service.

____ **16.** The elements needed to produce quality firefighting foam include:
   A. mechanical aeration, air, water, and concentrate.
   B. air, concentrate, eductor, and CO_2.
   C. proportioner, CO_2, and eductor.
   D. aspiration, subsurface injection, and air.

____ **17.** Alcohol resistant AFFF (Aqueous Film Forming Foam) can be utilized on hydrocarbon fires at _____ percent proportions.
   A. two
   B. three
   C. ten
   D. nine

____ **18.** Which of the following **is not** a polar solvent?
   A. Alcohol
   B. Acetone
   C. Kerosene
   D. Ketone

____ 19. For application of aqueous film-forming foam, eductors or proportioners operate on a(n) _____ principle.
   A. Bernoulli's
   B. Venturi
   C. induction
   D. pressure

____ 20. Foams in use currently are of the mechanical type and must be _____ and _____ before they can be used.
   A. proportioned, blended
   B. stirred, aerated
   C. mixed, proportioned
   D. proportioned, aerated

____ 21. Production of an adequate amount of bubbles to form an effective foam blanket is the definition of:
   A. proportioning
   B. aeration
   C. mixing
   D. blending

____ 22. To ensure maximum effectiveness, use foam concentrates _____ the specific percentage for which they are intended to be proportioned.
   A. within 10% greater than
   B. only at
   C. plus or minus 5% of
   D. within plus or minus 2% of

____ 23. Foams designed for Class B fuel fires are used at _____ percent concentration.
   A. 6 to 10
   B. 2 to 8
   C. 5 to 12
   D. 1 to 6

____ 24. Regarding the application of firefighting foam, the major difference between hydrocarbon and polar solvent fuels is that:
   A. hydrocarbons are miscible.
   B. polar solvents float on water.
   C. polar solvents are water soluble.
   D. Both A and C are correct.

____ 25. Fuel resistance is:
   A. the ability of the foam to stand up to the heat of the fire.
   B. how fast the foam spreads across the surface of a fuel.
   C. the ability to contain or control the production of fuel vapors.
   D. the ability to tolerate the fuel and to avoid being saturated.

___ **26.** Solubility is defined as:
   A. having the ability to mix with water.
   B. the study of fluids at rest and in motion.
   C. an aggregate of gas-filled bubbles formed from aqueous solutions.
   D. referring to evidence gathered at arson scenes.

___ **27.** The <u>preferred</u> method of controlling flammable liquid fires is:
   A. the use of foam.
   B. the use of large amounts of water.
   C. the use of unmanned nozzles.
   D. letting the fire burn undisturbed.

___ **28.** Water can be used at fires involving a flammable liquid as a cooling agent, mechanical tool, protective cover, and:
   A. mixing agent.
   B. substitute medium.
   C. vapor containment.
   D. wicking action.

___ **29.** The safest recommended means for a fire fighter to disconnect electrical service to a building is to:
   A. cut the service entrance wire.
   B. pull the meter.
   C. locate the nearest transformer and deactivate it.
   D. shut off the main power breaker/fuse in the panel box.

___ **30.** What type of construction has structural members (including walls, columns, beams, floors, and roofs) that are made of noncombustible or limited-combustible materials?
   A. Type I
   B. Type III
   C. Type IV
   D. Type V

___ **31.** Which of the following is a hazard associated with truss and lightweight construction?
   A. If one member fails, the entire truss is likely to fail.
   B. Once a truss fails, the one next to it is likely to fail.
   C. Trusses will begin to fail after five to ten minutes of exposure to fire.
   D. All of the above.

___ **32.** Which of the following <u>is not</u> an indicator of potential building collapse?
   A. Prolonged fire operations in the fire building
   B. Deteriorated mortar between the masonry
   C. Walls that appear to be leaning
   D. Large amounts of steam coming from ventilation openings

_____ 33. **Directions**: Read the following statements, then select your answer from alternatives A-D below.

To ensure that there is little danger of injury, a fire ax should be carried:
1. on the shoulder with the edge pointed toward the ground.
2. with the ax blade away from the body, or protected.
3. with pick-head axes, the pick should be covered with a hand.

   A. All three statements are true.
   B. Statement 1 and Statement 2 are false; Statement 3 is true.
   C. Statement 1 is true and Statements 2 and 3 are false.
   D. Statement 1 is false and Statements 2 and 3 are true.

_____ 34. What kind of heat energy is the heat of compression?
   A. Chemical
   B. Electrical
   C. Mechanical
   D. Nuclear

_____ 35. Natural roof ventilation openings consist of:
   A. cutting a 4 ft. X 4 ft. hole.
   B. trench cutting and smoke ejectors.
   C. scuttle hatches, skylights, and stairwell openings.
   D. fire streams from aerial ladders directed across a ventilation opening.

_____ 36. If a ventilation opening is made directly above a fire, it will tend to _____ the fire.
   A. spread
   B. mushroom
   C. localize
   D. extinguish

_____ 37. Which of the following is considered a natural or normal roof opening?
   A. Parapet
   B. Skylight
   C. Soffit
   D. Fire stop

_____ 38. Before cutting an opening in a roof, fire fighters should:
   A. inspect their cutting tools for sharpness.
   B. check for natural or existing openings.
   C. open all top windows on the windward side of the building.
   D. open all bottom windows on the leeward side of the building.

_____ 39. For purposes of ventilation, a large opening at least _____ feet is much better than several small ones.
   A. 1 X 1
   B. 2 X 2
   C. 3 X 3
   D. 4 X 4

____ **40.** Positive pressure ventilation is effective:
   A. when opening of doors and windows in the structure can be controlled.
   B. only if you can create a lower pressure zone in the structure.
   C. if the exhaust opening is smaller than the entry point, creating a Venturi effect.
   D. if an entire floor is ventilated at a time, starting at the highest floor and working down.

____ **41.** As a last resort, when ventilating a basement fire, _____ may be used to allow smoke and heat to escape.
   A. the HVAC system
   B. a hole cut in the floor above the fire
   C. a fire stream to break a window
   D. positive pressure force

____ **42.** In high-rise firefighting situations, typically the fire attack will be initiated from:
   A. the fire floor.
   B. one floor below the fire.
   C. one floor above the fire.
   D. two floors below the fire.

____ **43.** During fire situations below grade level, coupled with good ventilation techniques, fire fighters may put the nozzle on a _____ pattern to assist in gaining access to the fire.
   A. straight stream
   B. combination
   C. blitz
   D. wide fog

____ **44.** If there is adequate ventilation, steam from a water fog will _____ the toxic gases present.
   A. compress
   B. absorb
   C. displace
   D. contain

____ **45.** Construction featuring exterior walls and structural members that are noncombustible or of limited combustible materials without additional fire-resistant protection is Type _____ construction.
   A. I
   B. II
   C. III
   D. V

____ **46.** Tenders combined with _____ can efficiently provide large volumes of water to a fire ground operation.
   A. large-diameter hose
   B. automatic nozzles
   C. portable water tanks
   D. ladder trucks

_____ 47. A type of wood framing that has vertical channels going from floor to floor, allowing a fire to travel uninterrupted is a _____ frame construction.
   A. platform
   B. open
   C. balloon
   D. box

_____ 48. Which of the following gases is not produced in fires?
   A. Carbon monoxide
   B. Hydrogen cyanide
   C. Carbon dioxide
   D. Oxygen

_____ 49. A factor that will not influence the air currents in a structure fire is:
   A. a vertical vent opening.
   B. outside wind direction.
   C. horizontal ventilation opening.
   D. attic temperature.

_____ 50. When performing roof ventilation, the primary hole should be cut:
   A. directly over the fire, if possible.
   B. as far away from the fire as possible.
   C. directly in the fire fighters' escape path.
   D. as close to the edge of the roof as practical.

_____ 51. Which of the following factors must be taken into consideration when fighting a structure fire?
   A. Building construction
   B. Length of time the fire has been burning
   C. Occupancy type
   D. All of the above.

_____ 52. Which of the following statements is incorrect?
   A. The secondary search is the most dangerous.
   B. Searching a building is completed in two different operations-primary and secondary search.
   C. During primary search, the team is often ahead of attack lines and may be above the fire.
   D. The primary search takes place in a rapid but thorough manner in areas most likely to have victims.

_____ 53. Which of the following statements regarding trench cut is incorrect?
   A. The trench cut is an offensive action.
   B. The trench cut is 2-4 feet wide.
   C. It is not opened until the entire cut is complete.
   D. It is made in coordination with interior crews.

____ **54.** LPG is _____ than air.
  A. 1.5 times lighter
  B. 2 times lighter
  C. 1.5 times heavier
  D. .5 times heavier

____ **55.** The acronym "BLEVE" stands for Boiling Liquid:
  A. Exhausting Vapor Explosion
  B. Expanding and Venting Explosion
  C. Expanding Vapor Explosion
  D. Exhausting Vapor Expansion

____ **56.** A BLEVE:
  A. most commonly occurs when flames contact the relief valve.
  B. can occur when insufficient water is applied to keep the tank cool.
  C. is a slow deterioration of the tank.
  D. is a condition caused by consolidation of vaporization.

____ **57.** A tank with an "LPG" label on it would contain:
  A. liquid phosgene gas
  B. liquefied petroleum gas
  C. liquefied pressurized gas
  D. liquid petroleum gas

____ **58.** Firefighters should always remember how they gained entry into a building because:
  A. entry may have an effect on the behavior of the fire.
  B. the fire investigator may want to know how entry was made.
  C. you may need to use the opening in case of a rekindle.
  D. you should always exit the same way you entered.

____ **59.** Prior to the fire investigator's arrival, fire fighters should _____ any evidence found.
  A. tag and photograph
  B. protect and preserve
  C. collect and package
  D. isolate and remove

____ **60.** Following a fire of suspicious origin, the fire fighter should:
  A. carefully wash all burned articles with water to clean them off.
  B. remove any possible evidence to the outdoors where it can be properly tagged, identified, and photographed.
  C. have the building owner walk through the building to pick up valuables.
  D. leave suspected evidence where it is found.

____ **61.** Hidden fire can be checked by using a(n):
  A. detector for different levels of carbon monoxide and oxygen.
  B. Halligan tool to remove the whole wall.
  C. plaster tool from the opposite side of the wall.
  D. electronic/infrared heat sensor.

___ 62. Once a fire investigator has completed the work required in gathering all of the required evidence and information from a fire scene, a thorough _____ can be done.
   A. ventilation
   B. demobilization
   C. inventory
   D. overhaul

___ 63. Any vehicles in which a person is entrapped must be _____ to prevent them from shifting and inflicting more damage or injuries.
   A. stabilized
   B. hosed down
   C. removed
   D. left untouched

___ 64. Which of the following tools is essential for stabilizing a vehicle?
   A. Hydraulic jack
   B. Block and tackle
   C. High-pressure air bags
   D. Cribbing

___ 65. Which of the following devices is designed to provide the maximum in spinal immobilization?
   A. Stretcher
   B. Blanket
   C. Backboard
   D. Litter

___ 66. You are operating as a member of a rescue company at the scene of an auto extrication. The scene, vehicle, and the patient(s) are stabilized, all hazards are controlled, and access and disentanglement have been accomplished. The next step in the extrication process should be:
   A. termination and post-incident analysis.
   B. patient packaging and patient removal.
   C. transporting patient(s) to appropriate facilities.
   D. initiating extrication operations on another vehicle, if needed.

___ 67. The process of erecting materials such as wood panels, timber, or jacks to strengthen a wall or prevent further collapse is known as:
   A. shoring.
   B. cribbing.
   C. packing.
   D. supporting.

___ 68. The **most common** hazard encountered at a confined space incident is:
   A. an oxygen deficient atmosphere.
   B. engulfment.
   C. collapse of structure.
   D. entrapment.

_____ 69. When lifting rescued persons or heavy objects fire fighters should:
   A. use their back to lift, not their legs.
   B. use their legs to lift not, their back.
   C. try to twist and reach at the same time.
   D. squat and lift with shoulders and arms.

_____ 70. When using the Blanket Drag, patients should always be dragged:
   A. feet first.
   B. head first.
   C. sideways.
   D. Either B or C is correct

_____ 71. What is the <u>first</u> choice of action when involved in an ice rescue?
   A. Throwing a line or reaching with a pole
   B. Pushing a long ladder over the ice
   C. Wearing a PFD to walk onto the ice
   D. Paddling a light boat out to the victim

_____ 72. Rapid Intervention Crew is defined as:
   A. any combination of single resources assembled for an assignment.
   B. the designation for a set number of resources of the same type and kind.
   C. a company designated to search for and rescue trapped fire fighters.
   D. a designated group that is used for rapid knock down of wildland fires.

_____ 73. An example of a manually activated fire alarm component is a:
   A. thermally sensitive device.
   B. flame detector.
   C. water flow detector.
   D. pull station.

_____ 74. A special communications device which allows the hearing or speech impaired to communicate via telephone is known as a _____ system.
   A. commercial phone
   B. TDD/TTY text phone
   C. direct line
   D. wireless

_____ 75. Manual detection systems typically have two problems. The first is that many are local only, and the second is that:
   A. they are very technical to use.
   B. they are placed in closets to reduce false alarms.
   C. building occupants need a key to operate system.
   D. a person must be present, awake, and alert to discover the fire and activate the system.

_____ 76. Heat detectors:
   A. cannot be used as part of a suppression system.
   B. are slow to activate.
   C. are expensive to install and operate.
   D. are responsible for most false alarms.

_____ 77. Smoke detectors work primarily on the principles of photoelectricity and:
   A. rate of rise.
   B. fixed temperature.
   C. ionization.
   D. laser beam.

_____ 78. The purpose of the fire department connection to a sprinkler system is to:
   A. supplement the water supply while maintaining operational pressure.
   B. provide water, since most systems are dependent on the fire department for water supply.
   C. boost the water to upper stories, since most water pressure is not sufficient to supply water above the sixth floor.
   D. add water pressure to the system because normal water distribution is inadequate when less than three heads are activated.

_____ 79. _____ sprinkler systems should be used in buildings where piping may be subjected to freezing temperatures.
   A. Wet-pipe
   B. Dry-pipe
   C. Deluge
   D. Residential

_____ 80. The full name of a PIV is _____ valve.
   A. position-indicator valve
   B. post-indicator valve
   C. plant-indicator valve
   D. positive-indicator valve

_____ 81. The _____ system is equipped with all sprinkler heads of the open type.
   A. wet-pipe
   B. dry-pipe
   C. deluge
   D. preaction

_____ 82. A fire department connection to a sprinkler system enables fire fighters to:
   A. connect handlines for attacking the fire.
   B. drain water from the system.
   C. increase volume and pressure.
   D. test the system.

_____ 83. The full name of an OS&Y valve is _____ valve.
   A. open stem and yoke
   B. outside, shut, and yoke
   C. outside shield and yoke
   D. outside screw and yoke

84. Every sprinkler system should be equipped with a main control valve located between the:
  A. riser and the branches.
  B. source of water supply and the sprinkler system.
  C. cross mains and the riser.
  D. fire department connection and the riser.

85. The control valve for a sprinkler system may be located at the system or outside the building. This valve should always be a(n) _____ valve.
  A. check
  B. indicating-type
  C. inspector's test
  D. quarter-turn

86. Which of the following **should not** be used as a control valve for an automatic sprinkler system?
  A. Post indicator valve
  B. Wall post indicator valve
  C. Outside screw and yoke
  D. Gate valve

87. An OS&Y valve is open when the threads are:
  A. retracted.
  B. extended.
  C. reversed.
  D. crossed.

88. A riser is a _____ pipe that supplies a sprinkler system.
  A. lateral
  B. horizontal
  C. diagonal
  D. vertical

89. A sprinkler deflector converts a water stream into:
  A. spray.
  B. steam.
  C. a fine mist.
  D. a solid stream.

90. Residential sprinkler systems are wet- or dry-pipe systems. The recommended piping for these systems is constructed of:
  A. lead.
  B. ductile iron.
  C. brass
  D. steel, copper, or plastic.

_____ 91. Which of the following sprinkler system components is used to limit water flow to one direction?
  A. OS&Y valve
  B. Check valve
  C. Control valve
  D. Butterfly valve

_____ 92. _____ systems remain the most reliable of all fire protection devices.
  A. Foam
  B. Dry chemical
  C. Automatic sprinkler
  D. Standpipe

_____ 93. <u>Directions</u>: Read the statements below. Select your answer from alternatives A-D below.

  1. The typical rating for sprinkler head in light hazard occupancy is 165°F
  2. It is acceptable to use a pendant-type sprinkler head in the upright position.
  3. Sprinkler heads are color coded by their temperature ratings.

  A. All statements are true.
  B. Statement 1 is true; statements 2 and 3 are false.
  C. Statement 1 is false; statements 2 and 3 are true.
  D. Statement 2 is false; statements 1 and 3 are true.

_____ 94. Under normal circumstances, the air pressure gauge on a dry-pipe sprinkler system will read _____ the water pressure gauge.
  A. the same as
  B. higher than
  C. lower than
  D. almost double

_____ 95. A dry-pipe sprinkler system is equipped with two different types of gauges. The air-pressure gauge is located _____ the clapper valve and the water-pressure gauge is located _____ the clapper.
  A. to the left side; to the right side of
  B. above; below
  C. below; above
  D. to the right side; to the left side of

_____ 96. A floor-plan sketch or diagram of a structure shows:
  A. a detailed view of the construction features of each floor and fire protection features.
  B. an outline of each floor of the building, including walls, partitions, openings, fire protection features, and areas of life hazard.
  C. a cutaway view of a particular portion of a building in relation to streets and hydrants.
  D. the building and grounds as they are actually seen in depth by the eye.

_____ 97. A sketch depicting the general arrangement of a property in reference to streets, adjacent properties, and other important features is known as a:
   A. plot/site plan.
   B. sectional view.
   C. floor plan.
   D. blueprint.

_____ 98. In single family residences, _____ are an important part of an early warning system, in order to increase the chance of survival.
   A. signaling devices
   B. manual pull stations
   C. smoke detectors
   D. supervisory devices

_____ 99. Before conducting a pre-incident survey and inspection for the purpose of prefire planning, the occupant should be notified in advance so:
   A. the occupant can correct any fire hazards before the fire department arrives.
   B. there will be minimum inconvenience to the owner, occupant, and fire department.
   C. the owner can have fire extinguishers filled.
   D. someone will be at the location to fill out fire department forms.

_____ 100. A hazard that could produce or promote a fire that may cause a large loss of life or property damage **best** defines a(n) _____ hazard.
   A. special
   B. common
   C. expected
   D. target

_____ 101. Which of the following is not considered to be an alarm initiating device?
   A. Water detector
   B. Visible products-of-combustion detector
   C. Flame detector
   D. Invisible products-of-combustion detector

_____ 102. Which type of alarm initiating device contains a small portion of radioactive material in its sensing chamber?
   A. Water detector
   B. Visible products-of-combustion detector
   C. Flame detector
   D. Ionization detector

_____ 103. The _____ process requires fire fighters to become familiar with community structures.
   A. preventive
   B. code enforcement
   C. investigative
   D. survey/site visit

____104. A local protective signaling system is intended primarily to alert:
   A. the fire department dispatcher.
   B. a private alarm contractor.
   C. occupants of the protected area.
   D. the hotel desk clerk.

____105. Visible products of combustion are **best** detected by a(n) _____ detector.
   A. photoelectric
   B. ionization
   C. ultraviolet
   D. infrared

____106. A(n) _____ detector is one that responds to sudden changes in temperature.
   A. disk thermostat
   B. photoelectric
   C. ionization
   D. rate-of-rise

____107. When a fire hydrant receives water from two or more directions, it is said to have a _____ feed.
   A. compound
   B. circulating/loop
   C. compensating
   D. primary/distribution

____108. Water wells and springs are considered what type of water supply?
   A. Surface water supply
   B. Ground water supply
   C. Lake supply
   D. River supply

____109. Direct pumping water systems are those in which water:
   A. moves directly into the distribution system by gravity flow.
   B. is supplied directly into the distribution system from elevated storage tanks.
   C. is pumped directly into the distribution system with no elevated storage.
   D. is pumped directly through the distribution system back into the main water supply.

____110. In a water supply system, the size of the water mains from the largest to the smallest are:
   A. primary, distributor, secondary.
   B. distributor, secondary, primary.
   C. secondary, primary, distributor.
   D. primary, secondary, distributor.

____111. Which of the following violates the principle of a loop feed hydrant?
   A. Primary feeders
   B. Secondary feeders
   C. Interconnecting distributors
   D. Dead-end water mains

_____ 112. _____ pressure is defined as the normal pressure existing on a system before a flow hydrant is opened.
  A. Residual
  B. Static
  C. Velocity
  D. Flow

_____ 113. Residual pressure is defined as:
  A. stored potential energy available.
  B. that part of the total available pressure that is not used to overcome friction or gravity.
  C. forward velocity pressure at the point of discharge.
  D. the minimum pressure required in a residential area.

_____ 114. The forward velocity pressure at a discharge opening that is recorded by a pitot tube and gauge while water is flowing is known as _____ pressure.
  A. flow
  B. static
  C. normal operating
  D. residual

_____ 115. The two types of valves found in water supply distribution systems are:
  A. indicating and non-indicating.
  B. ball and check.
  C. in-line and flow.
  D. All of the above.

_____ 116. Which of the following is not one of the component parts of a dry-barrel fire hydrant?
  A. Operating stem
  B. Stem nut
  C. Post-indicator gate
  D. Drain hole

_____ 117. Which of the following is not a component of a grid system?
  A. Primary feeders
  B. Secondary feeders
  C. Distributors
  D. Risers

_____ 118. The part of the total pressure that is lost when water is forced through pipe, fittings, fire hose, and adapters is called:
  A. residual pressure.
  B. flow pressure.
  C. friction loss.
  D. static pressure.

_____ 119. The following figure depicts a _____ hydrant.
   A. dry-barrel
   B. wet-barrel
   C. drafting
   D. dry

_____ 120. A rate-of-rise detector will respond to a rapid change in temperature. This type of detector is:
   A. manually activated.
   B. a type of auxiliary protective signal.
   C. faster to react than a fixed temperature detector.
   D. located by the annunciator panel.

_____ 121. A _____ system is used to protect large commercial and industrial buildings and is staffed and supervised on the property.
   A. central station
   B. local protective
   C. remote
   D. proprietary

_____ 122. How does a deluge system differ from other types of sprinkler systems?
   A. It has air in the system until activated.
   B. It is installed in floors not ceilings.
   C. All sprinkler heads are open without any fusible heads.
   D. All of the above.

_____ 123. Which statement regarding residential sprinklers is <u>incorrect</u>?
   A. They are smaller but more expensive versions of wet- or dry-pipe sprinkler systems.
   B. The water supply is combined with the domestic system.
   C. The use of plastic pipe is allowed.
   D. They are designed to control the level of fire involvement such that residents can escape.

___124. In a sprinkler system, what is the function of the valve known as the backflow preventer or clapper?
   A. It prevents water from entering the system.
   B. It prevents water from system to reenter the public water supply.
   C. It prevents water from entering noninvolved areas.
   D. It protects system from water surges.

___125. Reinforced concrete has excellent _____ strength when it cures.
   A. shear
   B. compressive
   C. torsional
   D. tensile

___126. Masonry load-bearing walls are **most commonly** found in _____ construction.
   A. ordinary
   B. frame
   C. fire resistive
   D. noncombustible

___127. A connecting plate used in truss construction that can be made of flat steel stock, light gauge metal, or plywood is the definition of a:
   A. joint.
   B. gusset plate.
   C. column.
   D. joist.

___128. What is the simplest and **most effective** method of achieving the fire service goal of the preservation of life and property?
   A. Prevention
   B. Improved technology
   C. More fire fighters
   D. More fire stations

___129. During the preincident site visit/survey, what information should be obtained and documented?
   A. Built-in fire protection
   B. Construction type
   C. Structure size, height and number of stories
   D. All of the above.

___130. The four fundamental components of a modern water system are:
   A. source, mains, feeders, and risers.
   B. primary, secondary, standpipes, and subscriber connections.
   C. pipes, valves, hydrants, and pumps.
   D. source, means of moving, treatment plant, and distribution system.

___ **131.** A fire hydrant that receives water from only one direction is called a _____ hydrant.
  A. one-way
  B. steamer
  C. circulating-feed
  D. dead-end

___ **132.** The smaller internal grid arrangement of a water distribution system that feeds hydrants, as well as the domestic and commercial requirements, **best** describes:
  A. primary feeders.
  B. secondary feeders.
  C. distributors.
  D. grid network.

___ **133.** A detector that is usually electrically operated with a bimetallic strip of two metals that expand at different rates, eventually bending to touch a contact point and complete the alarm circuit is called a(n) _____ detector.
  A. rate-of-rise heat
  B. ionization smoke
  C. photoelectric smoke
  D. fixed-temperature heat

___ **134.** Which of the following statements is **incorrect**?
  A. Sprinkler systems are designed to automatically distribute water through sprinklers that are placed at set intervals on a system of piping to extinguish or control the spread of fires.
  B. Most sprinkler heads detect the heat of a fire and begin to apply water directly over the source of the heat.
  C. Sprinkler systems are highly ineffective.
  D. Sprinkler heads, unless deluge type heads, are heat sensitive-devices that react to a fixed temperature.

___ **135.** Large pipes that carry large quantities of water to various points along the water supply system for distribution to smaller mains **best** defines:
  A. primary feeders.
  B. secondary feeders.
  C. distributors.
  D. grid network.

___ **136.** A sprinkler head with a temperature rating of 135°-170°F would be:
  A. orange.
  B. red.
  C. white.
  D. either uncolored or black.

____137. Which of the following **is not** a type of sprinkler system?
   A. Preaction
   B. Dry pipe
   C. Deluge
   D. Total flooding

____138. The simplest sprinkler system in design and operation is the _____ system.
   A. wet pipe
   B. deluge
   C. dry pipe
   D. preaction

____139. During a fire, the control valve on a sprinkler system should be closed:
   A. as soon as the fire department arrives.
   B. after ensuring the fire is out or completely under control.
   C. prior to advancing hoselines into fire area.
   D. when building occupants decide fire is out.

____140. Dry pipe systems are used in all of the following incidents **except**:
   A. in buildings that refrigerate or freeze materials.
   B. in unheated buildings.
   C. outdoor applications where freezing temperatures occur.
   D. where rapid activation is required.

____141. A post indicator valve (PIV) is:
   A. a device to speed the operation of the dry pipe valve by detecting the decrease in air pressure.
   B. designed to control the head pressure at the outlet of a standpipe system to prevent excessive nozzle pressures in hoselines.
   C. a control valve that is mounted on a post case with a small window reading either "open" or "shut."
   D. a control valve that is mounted on a wall in a metal case with a small window reading either "open" or "shut."

____142. A record should be kept on each section of fire hose; information consisting of _____ should be recorded.
   A. the gallons per minute flowed during the test
   B. the annual number of fire responses at which it was used
   C. the date and results of the annual test
   D. number of threads per inch on each coupling

___143. The following figure depicts a _____ hydrant.
A. dry-barrel
B. wet-barrel
C. drafting
D. dry

___144. A network of intermediate-sized pipe that reinforces the overall grid system by forming loops that interlock primary feeders **best** defines:
A. primary loop.
B. secondary feeders.
C. distributors.
D. grid network.

___145. Which of the following pressures can be measured by a pitot tube?
A. Static
B. Normal operating
C. Residual
D. Flow

___146. A flow test from a fire hydrant will reveal:
A. a hydrant's coefficient of discharge.
B. how much flow pressure is available.
C. how much water will flow from any hydrant in the grid at any given time.
D. the size of the main on which that particular hydrant is installed.

___147. Directions: Read the statements below. Select your answer from alternatives A-D below.
1. Flow pressure is only measured through a nozzle.
2. Static pressure is stored potential energy available to move water through pipes, hoses, fittings, and adapters.
3. Residual pressure is that part of the total available pressure that **is not** used to overcome friction or gravity while forcing water through pipe, fittings, fire hose, and adapters.

A. Statements 1, 2, and 3 are true.
B. Statement 1 is true; statements 2 and 3 are false.
C. Statements 1 and 3 are false; statement 2 is true.
D. Statement 1 is false; statements 2 and 3 are true.

____**148.** When measuring flow pressure with a hand-held pitot gauge, the tube should be held:
    A. even with the end of the hydrant outlet.
    B. at a distance equal to one-half the diameter of the outlet.
    C. at a distance equal to the diameter of the outlet.
    D. approximately two inches from the end of the outlet.

____**149.** The term for taking water from a static source using fire department pumpers is:
    A. feeding.
    B. drafting.
    C. distributing.
    D. flowing.

____**150.** Failure to open a dry-barrel hydrant fully will result in a reduced amount of available water and will contribute to:
    A. sedimentation.
    B. susceptibility to freezing.
    C. difficulty in draining the main.
    D. ground erosion.

Did you score higher than 80% on Examination II-3? Circle Yes or No **in ink**.

Feedback Step

Now, what do we do with your Yes and No answers through the NFPA 1001 Examination Preparation process? First, return to any response that has No circled. Go back to the highlighted answers for those examination items missed and read and study the paragraph *preceding* the location of the answer, as well as the paragraph *following* the paragraph where the answer is located. This will expand your knowledge base for the missed question, put it in a broader perspective, and improve associative learning. Remember, we are trying to develop mastery of the required knowledge. Scoring 80% on an examination is good but it **is not** mastery performance. To come out in the top of your group, you must score well above 80% on your training, promotion, or certification examination.

Phases III and IV focus on getting you ready for the examination process by recommending activities that have a positive impact on the emotional and physical part of examination preparation. By evaluating your own progress through SAEP, you have determined that you have a high level of knowledge. Taking an examination for training, promotion, or certification is a competitive event. Just as in sports, **total preparation** is vitally important. Now you need to get all the elements of good preparation in place so that your next examination experience will be your best ever. Phase 3 is next! But first a review of Key Rules for Taking an Examination and Helpful Hints.

Summary of Key Rules for Taking an Examination

Rule 1—Examination preparation is not easy. Preparation is 95% perspiration and 5% inspiration.

Rule 2—Follow the steps very carefully. Do not try to reinvent or shortcut the system. It really works just as it was designed to!

Rule 3— Mark with an "X" any examination items for which you guessed the answer. For maximum return on effort, you should also research any answer that you guessed, even if you guessed correctly. Find the correct answer, highlight it, and then read the entire paragraph that contains the answer. Be honest and mark all questions you guessed. Some examinations have a correction for guessing built into the scoring process. The correction for guessing can reduce your final examination score. If you are guessing, you are not mastering the material.

Rule 4—Read questions twice if you have any misunderstanding and especially if the question contains complex directions or activities.

Rule 5—If you want someone to perform effectively and efficiently on the job, the training and testing program must be aligned to achieve this result.

Rule 6—When preparing examination items for job-specific requirements, the writer must be a subject matter expert with current experience at the level that the technical information is applied.

Rule 7—Good luck = Good preparation.

Helpful Hint

Studying the correct answers for missed items is a critical step in return on effort! The focus of attention is broadened and many times new knowledge is gained by expanding association and contextual learning. During our research and field test, self-study during this step of SAEP resulted in gain scores of 17 points between the first examination administered to the third examination. A gain score of 17 points can move you from the lower middle to the top of the list of persons taking a training, promotion, or certification examination. That is a competitive edge and prime example of return on effort in action. Remember: Maximum effort = Maximum results!

Summary of Helpful Hints

Helpful Hint—Most of the time your first impression is the best. More than 41% of <u>changed</u> answers during our SAEP field test were changed from a right answer to a wrong answer. Another 33% changed their answer from wrong to wrong. Only 26% of changed answers were from wrong to right. In fact, a number of changed answers resulted in three participants <u>not</u> making a perfect score of 100% when changing one right answer to a wrong one! Think twice before you change your answer. The odds don't appear to be in your favor.

Helpful Hint—Researching correct answers is one of the most important activities in SAEP. Locate the correct answer for all missed examination items. Highlight the correct answer. Then read the entire paragraph containing the answer. This will put the answer in context for you and provide important learning by association.

Helpful Hint—Proceed through all missed examination items using the same technique. Reading the entire paragraph improves retention of the information and helps you develop an association with the material and learn the correct answers. This step may sound simple. A major finding during the development and field testing of SAEP was that you learn from your mistakes.

Helpful Hint—Follow the steps carefully to realize the best return on effort. Would you consider investing your money in a venture without some chance of return on that investment? Examination preparation is no different. You are investing time expecting a significant return for that time. If, indeed, time is money, then you are investing money and are due a return on that investment.

Helpful Hint—Try to determine why you selected the wrong answer. Usually something influenced your selection. Focus on the difference between your wrong answer and the correct answer. Carefully read and study the entire paragraph containing the correct answer. Highlight the answer just as you did for Examination I-1.

Helpful Hint—Studying the correct answers for missed items is a critical step in return on effort! The focus of attention is broadened and many times new knowledge is gained by expanding association and contextual learning. During our research and field test, self study during this step of SAEP resulted in gain scores of 17 points between the first examination administered to the third examination. A gain score of 17 points can move you from the lower middle to the top of the list of persons taking a training, promotion, or certification examination. That is a competitive edge and prime example of return on effort in action. Remember: Maximum effort = Maximum results!

PHASE III

How Examination Developers Think – Getting Inside Their Heads

Now that you've finished the examination practice, this additional information will assist you in understanding and applying examination-taking skills. Developing your knowledge of how examination professionals think and prepare examinations is not cheating. Most serious examination takers have spent many hours reviewing various examinations to gain an insight into the technology. It is a demanding technology when used properly. You probably already know this if you have prepared examination items and administered them in your fire department. Phase III will not cover all the ways and means of examination-item writing. There are many techniques used by examination-item writers, far too many to cover in this book. The focus is on key techniques that will help you achieve a better score on your examination.

How Are Examination Items Derived?

Professional examination item writers use three basic techniques to derive examination items from text or technical reference materials. The most common technique is examination items taken verbatim from materials in the reference list. This technique doesn't work well for mastering information. The verbatim form of testing encourages rote learning or memorizing the material. The results of this type of learning are not long lasting, nor are they appropriate for learning and retaining critical knowledge you must have for on-the-job performance. The SAEP process does not base the majority of examination questions on NFPA 1001 using the verbatim technique. Professional examination item writers tend to use verbatim testing at the very basic level of job classifications. A Fire Fighter I, for instance, is expected to learn many basic facts. At this level, verbatim examination items can be justified.

In the higher ranks of Fire and Emergency Medical Services, other methods are more beneficial and productive for mastering higher cognitive knowledge and skills. A commonly used technique at the higher cognitive levels of an occupation, such as Fire Officer, will rely on other means. The most important technique at the higher cognitive levels is using deduction as the basis for examination items. This technique requires logic and analytical skills and often requires the examination taker to read materials several times to answer the examination item. It is not, then, a matter of the information that results in a verbatim answer. At the Fire Fighter I level, most activities are carefully supervised by a more experienced fire fighter or company officer. At this level, the fire fighter is expected to closely follow commands and is encouraged not to use deductive reasoning that can lead to "freelance" firefighting tactics. As one progresses to Fire Fighter II and gains experience, deductive reasoning and inferences skills are developed and applied. Most of the deductive reasoning and inference is related to personal safety and the safety of those on the scene. Most size up and strategies are developed and passed from the officers on the scene to the fire fighters.

Rule 5
If you want someone to perform effectively and efficiently on the job, the training and testing program must be aligned to achieve this result.

Rule #5 is paramount for fire fighters. Effective and efficient fire fighters are able to receive fireground commands, follow instructions, and perform their tasks safely and as rapidly as they can. There are limited opportunities for fire fighters to do much else since they are the first line of action at the emergency scene.

An example of deductive reasoning: an incident call is received from the Telecommunicator that an infant has a high temperature and is convulsing.

Just this amount of information should cause the first responder to immediately plan the response, conduct size up activity, and review infant care procedures in route. Some of these deductive responses will have you deal with the infant's age, past medical history, location, access, and many other possible deductions. If you have an EMT or Paramedic background, a list of several items could be deduced that would expedite an efficient and effective response to the incident.

You can probably think of many firefighting tasks and circumstances that rely on deductive reasoning. The more you gain experience on the fireground as a fire fighter the more you will get to practice deductive reasoning and inference from emergency data, the more efficient and effective you will become whether it is ventilating a roof or attending to the emergency needs of an infant.

Legendary Coach Vince Lombardi was once asked about the precision performance of his offensive and defensive teams. The comment was made that Lombardi must spend a lot of time on the practice field to get those results. Vince is quoted as saying, "Practice doesn't make perfect, only perfect practice makes perfect." This is exactly what is required to be an outstanding examination taker. Most people do not perfectly practice examination taking skills.

Another technique used by the professionals relies on inference or implied answers. Once again, the examination-item writer doesn't rely on verbatim techniques. Inference requires contrasting, comparing, analyzing, evaluating, and other high-level cognitive skills. Tables, charts, graphs and other instruments for presenting data provide an excellent means for deriving inference-based examination items. Implied answers are logic based. They rely on your ability to use logical processes or series of facts to arrive at a plausible answer. For example, recent data supplied by the National Fire Protection Association stated that heart attacks are still the leading cause of death for fire service personnel. Other data supplied stated that strains and sprains are still the leading cause of injuries on the job. Inference can be made from this relatively simple statement of information. A Safety Officer can infer to his own personnel and use the information as a trigger for checking on personnel, conducting surveys, reviewing accident records, and comparing the results with actual experience. Is our Fire Department doing better or worse in these important health issues? Are our fire and emergency service personnel getting the right exercise? Are we diligent in keeping the station and fire ground free from the activities that may lead to strains and sprains? The basic inference here is that your Fire Department is similar or different in some ways from the generalized data.

Knowing why it is sometimes difficult to find an answer to an examination item may be that it is measuring your ability to deduce and make inferences or to get the implied answer from the technical materials.

How Are Examination Items Written and Validated?

Once the information is identified and the technique for writing an examination item decided on, the professional will prepare a draft. The draft examination item is then

PHASE III, GOOD PRACTICES IN EXAMINATION ITEM AND EXAMINATION DEVELOPMENT

referenced to specific technical information such as a textbook, manufacturer's manual, or other related technical information. If the information is derived from a job-based requirement, then it should also be validated by job incumbents (i.e., those who are actually performing in the occupation at the specific level of the required knowledge).

Rule 6

When preparing examination items for job-specific requirements, the writer must be a subject matter expert with current experience at the level where the technical information is applied.

Rule #6 ensures that the examination item has a basic level of job content validity. The final level of job content validity is obtained by use of committees or surveys of job incumbents who certify the information to be current and required on the job. The information must be in a category of "need to know" or "must know" to be considered job relevant. The technical information must be accurate. Since subject matter experts do need basic training in examination-item writing, it is recommended that a professional in examination technology be part of the review process so that basic rules and guidelines of the industry are followed.

Last, but not least, the examination items must be field tested. Once this is done, statistical and analytical tools are available to help revise and improve the examination items. These techniques and tools go well beyond the scope of this examination prep book. Professionals are available to conduct these data analyses, and they should be used.

Good Practices in Examination Item and Examination Development

The most reliable examinations are objective. That is, they have only one answer that is accepted by members of the occupation. This objective quality permits fair and equitable examinations. The most popular objective examination items are the multiple choice, true/false, matching, and completion (fill in the blanks).

There are ten rules for valid and reliable job-relevant examinations in the Fire and Emergency Medical Services:

1. They do not contain trick questions.
2. They are short and easy to read, using language and terms appropriate to the target examination population.
3. They are supported with technical references, validation information, and data on their difficulty, discrimination, and other item analysis statistics.
4. They are formatted to meet recognized testing standards and examples.
5. They focus on the "need to know" and "must know" aspects of the job.
6. They are fair and objective.
7. They are not based on obscure and trivial knowledge and skills
8. They can be easily defended in terms of job content requirements.
9. They meet National and other professional job qualification standards.
10. They demonstrate their usefulness as part of a comprehensive testing program, including written, oral, and performance examination items.

The primary challenges of job-relevant examinations are their currency and validity. Careful recording of data, technical reference sources, and examination writer qualifications are important. Exams that affect someone's ability to be promoted, certified, or licensed, as well as to complete training that leads to a job, have exacting requirements in both published documents and in the laws of the land.

Three Common Myths of Examination Construction

1. If in doubt about the answer, select the longest answer in a multiple choice examination item.

    Myth - Professional examination-item writers use short answers as correct ones at an equal or higher percentage than longer answers. Remember, there are usually choices A-D. That leaves three other possibilities for the correct answer other than the longest. Statistically speaking the longest answer is less likely to be correct.

2. If in doubt about the answer in a multiple choice examination item, select "C".

    Myth - Computer technology and examination-item banking permit multiple versions of examinations to be simultaneously developed. This is typically achieved by moving the correct answer to different locations (for example, Version 1 will have the correct answer in the "C" position, Version 2 in the "D" position and so forth).

3. Watch for errors in singular examination-item stems with plural choices in either of the A-D answers or vice versa.

    Myth – Most computer-based programs have spelling and grammar checking utilities. If this mistake occurs, an editing error is the probable cause and usually has nothing to do with detecting the right answer.

Some Things That Work

1. Two to three days before your examination, review the examination items you missed in SAEP. Read those highlighted answers and the entire paragraph one more time.
2. During the examination, carefully read the examination item twice. Once you have selected your answer, read the examination item and answer together. This technique can key information that has been studied during your examination prep activities.
3. Apply what you learned in SAEP. Eliminate as many distracters as possible to improve the probability of getting the correct answer.
4. Pace yourself. Know how much time you have to take the examination. If an examination item is requiring too much time, write its number down and continue on to the next examination item. Often, an examination item later in the examination will key your memory and make the examination item easier to answer. (For a time pacing strategy see the Examination Pacing Table at the end of Phase IV)
5. Do not panic if there are examination items you do not know. Leave them to answer later. The most important thing is to finish the examination because there may be several examination items at the end of the examination that you know.
6. Once again, as time runs out for taking the examination, do not panic. Concentrate on answering those difficult examination items that you skipped.
7. Double check your answer sheets to make sure you have not accidentally left an answer blank.
8. Once you complete the examination, return to the difficult examination items. Often, while taking an examination, other examination items will cause you to remember or associate those answers with the difficult examination-item answers.

The longer the examination, the more likely you will be to gather information to answer the difficult examination items.

There are many other helpful hints that can be used. If you want to research the materials on how to take examinations and raise your final score, go to the local library, bookstore, or the Web for additional resources. The main reason we developed SAEP is to provide practice and help you develop examination taking skills you can use throughout your life.

PHASE IV

The Basics of Mental and Physical Preparation

Mental Preparation - I Can Get My Head Ready!

The two most common mental blocks to examination taking are examination anxiety and fear of failure. In the Fire and Emergency Medical Service these feelings can cause significant performance barriers. Severe conditions may require some professional psychological assistance, which is beyond the scope of this prep book.

The root cause of feelings of examination anxiety and fear of failure is often a lack of self confidence. SAEP was designed to help improve self confidence by providing evidence of mastery. Look at your scores as you progress through Phase I or Phase II. Review your Personal Progress Plotter located at the end of the Introduction Section. It will help you gain confidence in your knowledge of the NFPA 1001 Standard. Look at your Personal Progress Plotter the day before your scheduled examination and experience renewed confidence.

Let's examine the meaning of anxiety. Knowing what it is will help you deal with it at examination time. According to Webster, anxiety is "uneasiness and distress about future uncertainties." Many of us have real anxiety about taking examinations, and it is a natural response for some, often prefaced by such questions as: Am I ready for this? Do I have a good idea of what is going to be on the examination? Will I make the lowest score? Will John Doe score higher than me?

These questions and concerns are normal. Remember that hundreds of people have gone through SAEP with an average gain score of 17 points. The preparation process will help you maintain confidence. Once again, check the evidence in your Personal Progress Plotter to see what you have accomplished.

Fear, according to Webster, is "Alarm and agitation caused by the expectation or realization of danger." Let's deal with this normal reaction to examinations. First, analyze the degree of fear you may be experiencing several days before the examination date. Then focus on the positive experiences you had as you finished SAEP. Putting fear in perspective by using positives to eliminate or minimize it is a very important examination taking skill. The more you focus on your positive accomplishments in mastering the materials, the less fear you will experience.

If fear and anxiety persist, even after taking steps to build your confidence, you may want to get some professional assistance. Do it now! Don't wait until the week before the examination. There may be real issues that the professional can help you deal with to overcome these feelings. Hypnosis and other forms of treatment have been found to be very helpful. Consult with a professional expert.

Physical Preparation - Am I Really Ready?

This is the element that is probably most ignored in examination preparation. In the Fire and Emergency Medical Service, examinations are often given at locations away from home. If this is the case, you need to be especially careful of key physical concerns. More will be said about that later.

In general, following these helpful hints will help you concentrate, enhance your examination performance, and add points to your score.

1. Do not "cram" for the examination. This is most important because it was found to be first in importance during our field test of SAEP. Cramming results in test anxiety, adds confusion, and tends to lessen the effectiveness of examination taking skills you possess. Avoid cramming!
2. Get a normal night's rest. It may even be wise to take a day off before the examination to rest. Do not schedule an all night shift right before your examination.
3. Avoid excessive stimulants or medications that inhibit your thinking processes. Eat at least three well balanced meals before the day of the examination. It is a good practice to carry a balanced energy bar (not candy) and a bottle of water into the examination area. Examination anxiety and fear can cause a dry mouth, which can lead to further aggravation. Nibbling on the energy bar also has a settling effect and supplies some "brain food."
4. If the examination is at an out-of-town location, you should:
   - Avoid a "night out with friends." Lack of rest, partying, and fatigue are major examination performance killers.
   - Check your room carefully. Eliminate things that may aggravate you, interfere with your rest, or cause any discomfort. If the mattress is not good, pillows are horrible, or the room has an unpleasant odor, change rooms or even hotels.
   - Wake up in plenty of time to take a relaxing shower or soaking bath. Do not put yourself in a "rush" mode. Things should be carefully planned so that you arrive at the examination site ahead of time, calm, and collected.
5. Listen to the examination proctor. The proctor usually has rules that you must follow. Important instructions and directions are usually given. Ask clarifying questions immediately and listen to the response to questions raised by the other examination takers. Most examination environments are carefully controlled and may not permit questions you raise that are covered in the proctor's comments or about the technical content in the examination itself. Be attentive, focus, and succeed.
6. Remain calm and breathe. Pace yourself. Apply your examination taking skills learned during SAEP.
7. Remember the analogy of an examination as a competitive event. If you want the competitive edge, then carefully follow all Phases of SAEP. This process has yielded outstanding results and will do so for you.

Time Management During Examinations

The following table will help you develop a method for pacing yourself during an examination. You should get familiar with the table and be able to construct your own when you are in the examination room and getting ready to start the examination process. It will take a few minutes but it will make a tremendous contribution to your time management during the examination.

Here is how the table works. First you divide the examination time into 6 equal parts. If you have 3_ hours (210 minutes) for the examination, then each of the six time parts contains 35 minutes – 210 divided by 6 = 35 minutes. Now divide the number of examination items by 5. Example: The examination has 150 examination items – 150 divided by 5 = 30. Now, with the math done, we can set up a table that tells you approximately how many examination items you should answer in 35 minutes (the equal time divisions). You should be on or near examination item #30 at the end of the first 35 minutes and so forth. Notice we divided the examination items by 5 and the time by 6. This extra time block of 35 minutes is used to double check your answer sheet, focus on difficult questions, and to calm your nerves. This technique will work wonders for your stress level, and yes, it will improve your examination score.

Examination Pacing Table (150 and 100 Examination Items)

Time for Examination	Minutes for 6 Equal Time Parts	Number of Examination Items	Examination Items Per Time Part	Time for Examination Review
210 Minutes (3.5 Hours)	35 Minutes	150	30 (# of examination items to be answered)	35 Minutes (Chilling and Double-checking Examination)
150 Minutes (2.5 Hours)	25 Minutes	100	20 (# of examination items to be answered)	(Chilling and Double-checking Examination)

Using the Examination Pacing Table can be simplified by using the time/examination item variables as either may change in the real examination environment. For instance, if time is changed, adjust the ratio of time to answer the examination items in each of the five time blocks. If examination item numbers increase or decrease, adjust the number of examination items to be answered in the time blocks.

Some precautions when using this time management strategy:

1. Do not panic if you run a few minutes behind in each time block. It is a time management strategy and should not stress you while using it. Most people tend to pick up their pace as they move into the examination.

2. During the examination carefully mark or note examination items that you need to return to during your review time block. This will help you expedite your examination completion check.

3. Do not be afraid to ask for more time to complete your examination. In most cases, the time limit is flexible or should be.

4. Last, but not least, double check your answer sheet to make sure you didn't leave blank responses and that you didn't double mark answers. Double markings are most often counted as wrong answers. Make sure any erasers are made cleanly. **Caution**: Make sure when you change your answer that you really want to do that. Odds are not in your favor unless something on the examination really influenced the change.

APPENDIX A

Examination I-1 Answer Key

Directions
Follow these steps carefully for completing the feedback step of Systematic Approach to Exam Preparation (SAEP):

1. After entering your scores, look up the answers for the examination items you missed as well as those you guessed, even if you guessed correctly. If you are guessing, it means the answer is not perfectly clear. In this process we are committed to making you as knowledgeable as possible.
2. Enter the number of missed and guessed examination items in the blank on the personal progress plotter.
3. Highlight the answer in the reference materials. Once you have highlighted the answer, read the paragraph preceding and the paragraph following the one in which the correct answer is located. Now that you have highlighted the answer, enter the paragraph number and page number next to the guessed or missed examination item on your examination. Count any part of a paragraph at the beginning of the page as one paragraph until you reach the paragraph containing your highlighted answer. This step will help you locate and review your missed and guessed examination items later in the process. This step is <u>essential</u> to learning the material in context and by association. These learning techniques (context/association) are the very backbone of the SAEP approach.
4. Once you have completed the feedback step, you may proceed to the next examination.

1. Reference: NFPA 1001, 5.1.1.1
   Delmar, *Firefighter's Handbook*, 2nd Edition, 1st Printing, page 607.
   IFSTA, *Essentials of Fire Fighting*, 4th Edition, 1st Printing, page 13.
   Jones and Bartlett, *Fundamentals of Fire Fighter Skills*, 1st Edition, 1st Printing, page 291.
   Answer: C

2. Reference: NFPA 1001, 5.1.1.1
   Delmar, *Firefighter's Handbook*, 2nd Edition, 1st Printing, page 35.
   IFSTA, *Essentials of Fire Fighting*, 4th Edition, 1st Printing, pages 12-13.
   Jones and Bartlett, *Fundamentals of Fire Fighter Skills*, 1st Edition, 1st Printing, pages 12-13.
   Answer: A

3. Reference: NFPA 1001, 5.1.1.1
Delmar, *Firefighter's Handbook*, 2nd Edition, 1st Printing, pages 35 and 607.
IFSTA, *Essentials of Fire Fighting*, 4th Edition, 1st Printing, page 13.
Jones and Bartlett, *Fundamentals of Fire Fighter Skills*, 1st Edition, 1st Printing, pages 12-13 and 291.
Answer: D

4. Reference: NFPA 1001, 5.1.1.1
Delmar, *Firefighter's Handbook*, 2nd Edition, 1st Printing, page 467.
IFSTA, *Essentials of Fire Fighting*, 4th Edition, 1st Printing, page 25.
Jones and Bartlett, *Fundamentals of Fire Fighter Skills*, 1st Edition, 1st Printing, pages 345 and 386.
Answer: A

5. Reference: NFPA 1001, 5.1.1.1
Delmar, *Firefighter's Handbook*, 2nd Edition, 1st Printing, page 40.
IFSTA, *Essentials of Fire Fighting*, 4th Edition, 1st Printing, page 15.
Jones and Bartlett, *Fundamentals of Fire Fighter Skills*, 1st Edition, 1st Printing, page 111.
Answer: A

6. Reference: NFPA 1001, 5.1.1.1
Delmar, *Firefighter's Handbook*, 2nd Edition, 1st Printing, pages 26 & 27.
IFSTA, *Essentials of Fire Fighting*, 4th Edition, 1st Printing, page 8.
Jones and Bartlett, *Fundamentals of Fire Fighter Skills*, 1st Edition, 1st Printing, pages 11-13.
Answer: D

7. Reference: NFPA 1001, 5.1.1.1
Delmar, *Firefighter's Handbook*, 2nd Edition, 1st Printing, page 40.
IFSTA, *Essentials of Fire Fighting*, 4th Edition, 1st Printing, page 15.
Jones and Bartlett, *Fundamentals of Fire Fighter Skills*, 1st Edition, 1st Printing, page 109.
Answer: C

8. Reference: NFPA 1001, 5.1.1.1
Delmar, *Firefighter's Handbook*, 2nd Edition, 1st Printing, page 29.
IFSTA, *Essentials of Fire Fighting*, 4th Edition, 1st Printing, page 8.
Jones and Bartlett, *Fundamentals of Fire Fighter Skills*, 1st Edition, 1st Printing, page 9.
Answer: A

9. Reference: NFPA 1001, 5.1.1.1, and 5.1.1.2
Delmar, *Firefighter's Handbook*, 2nd Edition, 1st Printing, page 427.
IFSTA, *Essentials of Fire Fighting*, 4th Edition, 1st Printing, pages 154, 155, and 165.
Jones and Bartlett, *Fundamentals of Fire Fighter Skills*, 1st Edition, 1st Printing, page 257.
Answer: A

10. Reference: NFPA 1001, 5.1.1.1, and 5.1.1.2
    Delmar, *Firefighter's Handbook*, 2nd Edition, 1st Printing, page 434.
    IFSTA, *Essentials of Fire Fighting*, 4th Edition, 1st Printing, pages 156 and 166.
    Jones and Bartlett, *Fundamentals of Fire Fighter Skills*, 1st Edition, 1st Printing, page 260.
    Answer: D

11. Reference: NFPA 1001, 5.1.1.1, and 5.1.1.2
    Delmar, *Firefighter's Handbook*, 2nd Edition, 1st Printing, page 425.
    IFSTA, *Essentials of Fire Fighting*, 4th Edition, 1st Printing, page 154.
    Jones and Bartlett, *Fundamentals of Fire Fighter Skills*, 1st Edition, 1st Printing, page 254.
    Answer: C

12. Reference: NFPA 1001, 5.1.1.1, and 5.1.1.2
    Delmar, *Firefighter's Handbook*, 2nd Edition, 1st Printing, pages 423-424.
    IFSTA, *Essentials of Fire Fighting*, 4th Edition, 1st Printing, page 171.
    Jones and Bartlett, *Fundamentals of Fire Fighter Skills*, 1st Edition, 1st Printing, page 242.
    Answer: A

13. Reference: NFPA 1001, 5.1.1.1, and 5.1.1.2
    Delmar, *Firefighter's Handbook*, 2nd Edition, 1st Printing, page 429.
    IFSTA, *Essentials of Fire Fighting*, 4th Edition, 1st Printing, page 156 and 169.
    Jones and Bartlett, *Fundamentals of Fire Fighter Skills*, 1st Edition, 1st Printing, page 262.
    Answer: C

14. Reference: NFPA 1001, 5.1.1.1, and 5.1.1.2
    Delmar, *Firefighter's Handbook*, 2nd Edition, 1st Printing, page 432.
    IFSTA, *Essentials of Fire Fighting*, 4th Edition, 1st Printing, pages 155 and 163.
    Jones and Bartlett, *Fundamentals of Fire Fighter Skills*, 1st Edition, 1st Printing, page 264.
    Answer: A

15. Reference: NFPA 1001, 5.1.1.1, and 5.1.1.2
    Delmar, *Firefighter's Handbook*, 2nd Edition, 1st Printing, pages 427-429.
    IFSTA, *Essentials of Fire Fighting*, 4th Edition, 1st Printing, pages 157, 170, and 172.
    Jones and Bartlett, *Fundamentals of Fire Fighter Skills*, 1st Edition, 1st Printing, page 257.
    Answer: C

16. Reference: NFPA 1001, 5.1.1.1, and 5.1.1.2
    Delmar, *Firefighter's Handbook*, 2nd Edition, 1st Printing, page 433.
    IFSTA, *Essentials of Fire Fighting*, 4th Edition, 1st Printing, page 156.
    Jones and Bartlett, *Fundamentals of Fire Fighter Skills*, 1st Edition, 1st Printing, pages 260-262.
    Answer: C

17. Reference: NFPA 1001, 5.1.1.1, and 5.1.1.2
Delmar, *Firefighter's Handbook*, 2nd Edition, 1st Printing, page 430.
IFSTA, *Essentials of Fire Fighting*, 4th Edition, 1st Printing, pages 156 and 169.
Jones and Bartlett, *Fundamentals of Fire Fighter Skills*, 1st Edition, 1st Printing, page 265.
Answer: B

18. Reference: NFPA 1001, 5.1.1.1, and 5.1.1.2
Delmar, *Firefighter's Handbook*, 2nd Edition, 1st Printing, page 455.
IFSTA, *Essentials of Fire Fighting*, 4th Edition, 1st Printing, page 157.
Jones and Bartlett, *Fundamentals of Fire Fighter Skills*, 1st Edition, 1st Printing, page 268.
Answer: C

19. Reference: NFPA 1001, 5.1.1.1 and 5.1.1.2
Delmar, *Firefighter's Handbook*, 2nd Edition, 1st Printing, page 429.
IFSTA, *Essentials of Fire Fighting*, 4th Edition, 1st Printing, pages 154 and 165.
Jones and Bartlett, *Fundamentals of Fire Fighter Skills*, 1st Edition, 1st Printing, page 259.
Answer: C

20. Reference: NFPA 1001, 5.2.1 and 5.2.1(A)(B)
Delmar, *Firefighter's Handbook*, 2nd Edition, 1st Printing, page 65.
IFSTA, *Essentials of Fire Fighting*, 4th Edition, 1st Printing, page 647.
Jones and Bartlett, *Fundamentals of Fire Fighter Skills*, 1st Edition, 1st Printing, page 87.
Answer: B

21. Reference: NFPA 1001, 5.2.1 and 5.2.1(A)(B)
Delmar, *Firefighter's Handbook*, 2nd Edition, 1st Printing, page 63.
IFSTA, *Essentials of Fire Fighting*, 4th Edition, 1st Printing, page 645.
Jones and Bartlett, *Fundamentals of Fire Fighter Skills*, 1st Edition, 1st Printing, page 83.
Answer: C

22. Reference: NFPA 1001, 5.2.1 and 5.2.1(A)(B)
Delmar, *Firefighter's Handbook*, 2nd Edition, 1st Printing, pages 51 and 71.
IFSTA, *Essentials of Fire Fighting*, 4th Edition, 1st Printing, page 638.
Jones and Bartlett, *Fundamentals of Fire Fighter Skills*, 1st Edition, 1st Printing, pages 74 and 96.
Answer: A

23. Reference: NFPA 1001, 5.2.1 and 5.2.1(A)(B)
Delmar, *Firefighter's Handbook*, 2nd Edition, 1st Printing, page 51.
IFSTA, *Essentials of Fire Fighting*, 4th Edition, 1st Printing, page 635.
Jones and Bartlett, *Fundamentals of Fire Fighter Skills*, 1st Edition, 1st Printing, pages 77, 94-95.
Answer: C

APPENDIX A, EXAMINATION I-1 ANSWER KEY

24. Reference: NFPA 1001, 5.2.1 and 5.2.1(A)(B)
Delmar, *Firefighter's Handbook*, 2nd Edition, 1st Printing, pages 51-52.
IFSTA, *Essentials of Fire Fighting*, 4th Edition, 1st Printing, page 635.
Jones and Bartlett, *Fundamentals of Fire Fighter Skills*, 1st Edition, 1st Printing, pages 94-95.
Answer: B

25. Reference: NFPA 1001, 5.2.3 and 5.2.3(A)(B)
Delmar, *Firefighter's Handbook*, 2nd Edition, 1st Printing, pages 65-66.
IFSTA, *Essentials of Fire Fighting*, 4th Edition, 1st Printing, page 647.
Jones and Bartlett, *Fundamentals of Fire Fighter Skills*, 1st Edition, 1st Printing, page 85.
Answer: D

26. Reference: NFPA 1001, 5.3.1 and 5.3.1(A)(B)
Delmar, *Firefighter's Handbook*, 2nd Edition, 1st Printing, page 174.
IFSTA, *Essentials of Fire Fighting*, 4th Edition, 1st Printing, page 121.
Jones and Bartlett, *Fundamentals of Fire Fighter Skills*, 1st Edition, 1st Printing, page 62.
Answer: B

27. Reference: NFPA 1001, 5.3.1 and 5.3.1(A)(B)
Delmar, *Firefighter's Handbook*, 2nd Edition, 1st Printing, pages 164-166.
IFSTA, *Essentials of Fire Fighting*, 4th Edition, 1st Printing, page 103.
Jones and Bartlett, *Fundamentals of Fire Fighter Skills*, 1st Edition, 1st Printing, pages 53-56.
Answer: B

28. Reference: NFPA 1001, 5.3.1 and 5.3.1(A)(B)
Delmar, *Firefighter's Handbook*, 2nd Edition, 1st Printing, page 156.
IFSTA, *Essentials of Fire Fighting*, 4th Edition, 1st Printing, pages 103-104, 107-108, and 176.
Jones and Bartlett, *Fundamentals of Fire Fighter Skills*, 1st Edition, 1st Printing, page 48.
Answer: C

29. Reference: NFPA 1001, 5.3.1, 5.3.1(A)(B), 5.3.5(A)(B), and 5.3.9(A)(B)
Delmar, *Firefighter's Handbook*, 2nd Edition, 1st Printing, page 156.
IFSTA, *Essentials of Fire Fighting*, 4th Edition, 1st Printing, page 97.
Jones and Bartlett, *Fundamentals of Fire Fighter Skills*, 1st Edition, 1st Printing, page 48.
Answer: B

30. Reference: NFPA 1001, 5.3.1, 5.3.1(A), 5.3.5(A)(B), 5.3.10(A), and 5.3.11(A)
Delmar, *Firefighter's Handbook*, 2nd Edition, 1st Printing, pages 145-146.
IFSTA, *Essentials of Fire Fighting*, 4th Edition, 1st Printing, pages 87-88.
Jones and Bartlett, *Fundamentals of Fire Fighter Skills*, 1st Edition, 1st Printing, pages 42-44.
Answer: D

31. Reference: NFPA 1001, 5.3.1, 5.3.1(A)(B), and 5.1.1.2
Delmar, *Firefighter's Handbook*, 2nd Edition, 1st Printing, pages 166-167.
IFSTA, *Essentials of Fire Fighting*, 4th Edition, 1st Printing, page 104.
Jones and Bartlett, *Fundamentals of Fire Fighter Skills*, 1st Edition, 1st Printing, pages 56 and 58.
Answer: A

32. Reference: NFPA 1001, 5.3.1, 5.3.1(A)(B), and 5.3.5(A)(B)
Delmar, *Firefighter's Handbook*, 2nd Edition, 1st Printing, page 145.
IFSTA, *Essentials of Fire Fighting*, 4th Edition, 1st Printing, page 88.
Jones and Bartlett, *Fundamentals of Fire Fighter Skills*, 1st Edition, 1st Printing, page 43.
Answer: D

33. Reference: NFPA 1001, 5.3.1, 5.3.1(A)(B), 5.3.5(A)(B), 5.3.10(A), and 5.3.11(A)
Delmar, *Firefighter's Handbook*, 2nd Edition, 1st Printing, page 145.
IFSTA, *Essentials of Fire Fighting*, 4th Edition, 1st Printing, pages 87-88.
Jones and Bartlett, *Fundamentals of Fire Fighter Skills*, 1st Edition, 1st Printing, page 43.
Answer: C

34. Reference: NFPA 1001, 5.3.1, 5.3.1(A)(B), and 5.3.5(A)(B)
Delmar, *Firefighter's Handbook*, 2nd Edition, 1st Printing, page 145.
IFSTA, *Essentials of Fire Fighting*, 4th Edition, 1st Printing, pages 87-88.
Jones and Bartlett, *Fundamentals of Fire Fighter Skills*, 1st Edition, 1st Printing, pages 42-44.
Answer: A

35. Reference: NFPA 1001, 5.3.1, 5.3.1(A)(B), and 5.3.5(A)(B)
Delmar, *Firefighter's Handbook*, 2nd Edition, 1st Printing, page 156.
IFSTA, *Essentials of Fire Fighting*, 4th Edition, 1st Printing, pages 95-96.
Jones and Bartlett, *Fundamentals of Fire Fighter Skills*, 1st Edition, 1st Printing, page 48.
Answer: A

36. Reference: NFPA 1001, 5.3.1, 5.3.1(A), 5.3.5(A)(B), 5.3.9, and 5.3.9(A)(B)
Delmar, *Firefighter's Handbook*, 2nd Edition, 1st Printing, pages 145 and 498.
IFSTA, *Essentials of Fire Fighting*, 4th Edition, 1st Printing, pages 91-92.
Jones and Bartlett, *Fundamentals of Fire Fighter Skills*, 1st Edition, 1st Printing, pages 42-44.
Answer: D

37. Reference: NFPA 1001, 5.3.1, 5.3.1(A)(B), 5.3.5, 5.3.5(A)(B), 5.3.9 and 5.3.9(A)(B)
Delmar, *Firefighter's Handbook*, 2nd Edition, 1st Printing, page 171.
IFSTA, *Essentials of Fire Fighting*, 4th Edition, 1st Printing, page 108.
Jones and Bartlett, *Fundamentals of Fire Fighter Skills*, 1st Edition, 1st Printing, page 58.
Answer: D

38. Reference: NFPA 1001, 5.3.1 and 5.3.1(A)(B)
Delmar, *Firefighter's Handbook*, 2nd Edition, 1st Printing, page 156.
IFSTA, *Essentials of Fire Fighting*, 4th Edition, 1st Printing, page 97.
Jones and Bartlett, *Fundamentals of Fire Fighter Skills*, 1st Edition, 1st Printing, page 48.
Answer: B

39. Reference: NFPA 1001, 5.3.1, 5.3.1(A), and 5.3.5(A)(B)
Delmar, *Firefighter's Handbook*, 2nd Edition, 1st Printing, page 156.
IFSTA, *Essentials of Fire Fighting*, 4th Edition, 1st Printing, page 97.
Jones and Bartlett, *Fundamentals of Fire Fighter Skills*, 1st Edition, 1st Printing, page 48.
Answer: C

40. Reference: NFPA 1001, 5.3.1, 5.3.1(A)(B), 5.3.5 and 5.3.5(A)(B)
Delmar, *Firefighter's Handbook*, 2nd Edition, 1st Printing, page 156.
IFSTA, *Essentials of Fire Fighting*, 4th Edition, 1st Printing, pages 96-97.
Jones and Bartlett, *Fundamentals of Fire Fighter Skills*, 1st Edition, 1st Printing, page 48.
Answer: C

41. Reference: NFPA 1001, 5.3.1, 5.3.1(A), 5.3.5, 5.3.5(A)(B), 5.3.9, 5.3.9(A)(B), 5.3.10, 5.3.10(A), 5.3.11 and 5.3.11(A)
Delmar, *Firefighter's Handbook*, 1st Edition, 1st Printing, pages 146-147.
IFSTA, *Essentials of Fire Fighting*, 4th Edition, 1st Printing, pages 88-89.
Jones and Bartlett, *Fundamentals of Fire Fighter Skills*, 1st Edition, 1st Printing, page 43.
Answer: C

42. Reference: NFPA 1001, 5.3.1, 5.3.1(A), 5.3.5, 5.3.5(A)(B), 5.3.9, 5.3.9(A)(B), 5.3.10, 5.3.10(A), 5.3.11 and 5.3.11(A)
Delmar, *Firefighter's Handbook*, 2nd Edition, 1st Printing, pages 151 and 158.
IFSTA, *Essentials of Fire Fighting*, 4th Edition, 1st Printing, pages 88-89.
Jones and Bartlett, *Fundamentals of Fire Fighter Skills*, 1st Edition, 1st Printing, page 45.
Answer: B

43. Reference: NFPA 1001, 5.3.2 and 5.3.2(A)(B)
Delmar, *Firefighter's Handbook*, 2nd Edition, 1st Printing, page 134.
IFSTA, *Essentials of Fire Fighting*, 4th Edition, 1st Printing, page 23.
Jones and Bartlett, *Fundamentals of Fire Fighter Skills*, 1st Edition, 1st Printing, pages 282-283.
Answer: C

44. Reference: NFPA 1001, 5.3.2 and 5.3.2(A)(B)
Delmar, *Firefighter's Handbook*, 2nd Edition, 1st Printing, page 120.
IFSTA, *Essentials of Fire Fighting*, 4th Edition, 1st Printing, page 23.
Jones and Bartlett, *Fundamentals of Fire Fighter Skills*, 1st Edition, 1st Printing, pages 282-283.
Answer: D

45. Reference: NFPA 1001, 5.3.2 and 5.3.2(A)(B)
Delmar, *Firefighter's Handbook*, 2nd Edition, 1st Printing, page 125.
IFSTA, *Essentials of Fire Fighting*, 4th Edition, 1st Printing, page 81.
Jones and Bartlett, *Fundamentals of Fire Fighter Skills*, 1st Edition, 1st Printing, pages 37-38.
Answer: B

46. Reference: NFPA 1001, 5.3.2 and 5.3.2(A)
Delmar, *Firefighter's Handbook*, 2nd Edition, 1st Printing, page 136.
IFSTA, *Essentials of Fire Fighting*, 4th Edition, 1st Printing, page 87.
Jones and Bartlett, *Fundamentals of Fire Fighter Skills*, 1st Edition, 1st Printing, page 31.
Answer: B

47. Reference: NFPA 1001, 5.3.2 and 5.3.2(A)
Delmar, *Firefighter's Handbook*, 2nd Edition, 1st Printing, page 128.
IFSTA, *Essentials of Fire Fighting*, 4th Edition, 1st Printing, page 83.
Jones and Bartlett, *Fundamentals of Fire Fighter Skills*, 1st Edition, 1st Printing, page 32.
Answer: B

48. Reference: NFPA 1001, 5.3.2 and 5.3.2(A)(B)
Delmar, *Firefighter's Handbook*, 2nd Edition, 1st Printing, page 129.
IFSTA, *Essentials of Fire Fighting*, 4th Edition, 1st Printing, page 84.
Jones and Bartlett, *Fundamentals of Fire Fighter Skills*, 1st Edition, 1st Printing, page 35.
Answer: C

49. Reference: NFPA 1001, 5.3.2 and 5.3.2(A)(B)
Delmar, *Firefighter's Handbook*, 2nd Edition, 1st Printing, page 135.
IFSTA, *Essentials of Fire Fighting*, 4th Edition, 1st Printing, page 100.
Jones and Bartlett, *Fundamentals of Fire Fighter Skills*, 1st Edition, 1st Printing, pages 35-37.
Answer: A

50. Reference: NFPA 1001, 5.3.3 and 5.3.3(A)(B)
IFSTA, *Essentials of Fire Fighting*, 4th Edition, 1st Printing, page 540.
Jones and Bartlett, *Fundamentals of Fire Fighter Skills*, 1st Edition, 1st Printing, page 28.
Answer: D

51. Reference: NFPA 1001, 5.3.4 and 5.3.4(A)(B)
Delmar, *Firefighter's Handbook*, 2nd Edition, 1st Printing, pages 512 and 515.
IFSTA, *Essentials of Fire Fighting*, 4th Edition, 1st Printing, pages 234 and 239.
Jones and Bartlett, *Fundamentals of Fire Fighter Skills*, 1st Edition, 1st Printing, pages 224 and 300.
Answer: A

52. Reference: NFPA 1001, 5.3.4 and 5.3.4(A)(B)
Delmar, *Firefighter's Handbook*, 2nd Edition, 1st Printing, page 523.
IFSTA, *Essentials of Fire Fighting*, 4th Edition, 1st Printing, page 255.
Jones and Bartlett, *Fundamentals of Fire Fighter Skills*, 1st Edition, 1st Printing, page 312.
Answer: B

53. Reference: NFPA 5.3.4 and 5.3.4(A)
Delmar, *Firefighter's Handbook*, 2nd Edition, 1st Printing, page 524.
IFSTA, *Essentials of Fire Fighting*, 4th Edition, 1st Printing, page 246.
Jones and Bartlett, *Fundamentals of Fire Fighter Skills*, 1st Edition, 1st Printing, page 307.
Answer: C

54. Reference: NFPA 5.3.4 and 5.3.4(A)
Delmar, *Firefighter's Handbook*, 2nd Edition, 1st Printing, page 522.
IFSTA, *Essentials of Fire Fighting*, 4th Edition, 1st Printing, page 245.
Jones and Bartlett, *Fundamentals of Fire Fighter Skills*, 1st Edition, 1st Printing, page 304.
Answer: A

55. Reference: NFPA 5.3.4 and 5.3.4(A)(B)
Delmar, *Firefighter's Handbook*, 2nd Edition, 1st Printing, page 513.
IFSTA, *Essentials of Fire Fighting*, 4th Edition, 1st Printing, page 265.
Jones and Bartlett, *Fundamentals of Fire Fighter Skills*, 1st Edition, 1st Printing, page 300.
Answer: C

56. Reference: NFPA 5.3.4 and 5.3.4(A)(B)
Delmar, *Firefighter's Handbook*, 2nd Edition, 1st Printing, page 511.
IFSTA, *Essentials of Fire Fighting*, 4th Edition, 1st Printing, page 243.
Jones and Bartlett, *Fundamentals of Fire Fighter Skills*, 1st Edition, 1st Printing, page 299.
Answer: B

57. Reference: NFPA 5.3.4 and 5.3.4(A)
Delmar, *Firefighter's Handbook*, 2nd Edition, 1st Printing, page 518.
IFSTA, *Essentials of Fire Fighting*, 4th Edition, 1st Printing, page 256.
Jones and Bartlett, *Fundamentals of Fire Fighter Skills*, 1st Edition, 1st Printing, pages 230 and 303.
Answer: C

58. Reference: NFPA 1001, 5.3.5, 5.3.5(A)(B), 5.3.9, and 5.3.9(A)(B)
Delmar, *Firefighter's Handbook*, 2nd Edition, 1st Printing, page 461.
IFSTA, *Essentials of Fire Fighting*, 4th Edition, 1st Printing, page 183.
Jones and Bartlett, *Fundamentals of Fire Fighter Skills*, 1st Edition, 1st Printing, pages 382 and 540.
Answer: B

59. Reference: NFPA 1001, 5.3.5, 5.3.5(A)(B), 5.3.9, and 5.3.9(A)(B)
Delmar, *Firefighter's Handbook*, 2nd Edition, 1st Printing, page 462.
IFSTA, *Essentials of Fire Fighting*, 4th Edition, 1st Printing, page 183.
Jones and Bartlett, *Fundamentals of Fire Fighter Skills*, 1st Edition, 1st Printing, page 540.
Answer: C

60. Reference: NFPA 1001, 5.3.5, 5.3.5(A)(B), 5.3.9, and 5.3.9(A)(B)
Delmar, *Firefighter's Handbook*, 2nd Edition, 1st Printing, pages 461-462.
IFSTA, *Essentials of Fire Fighting*, 4th Edition, 1st Printing, page 180.
Jones and Bartlett, *Fundamentals of Fire Fighter Skills*, 1st Edition, 1st Printing, page 382.
Answer: D

61. Reference: NFPA 1001, 5.3.5 and 5.3.5(A)(B)
Delmar, *Firefighter's Handbook*, 2nd Edition, 1st Printing, page 725.
IFSTA, *Essentials of Fire Fighting*, 4th Edition, 1st Printing, pages 29-30.
Jones and Bartlett, *Fundamentals of Fire Fighter Skills*, 1st Edition, 1st Printing, pages 285 and 540.
Answer: B

62. References: NFPA 1001, 5.3.6 and 5.3.6(A)(B)
Delmar, *Firefighter's Handbook*, 2nd Edition, 1st Printing, pages 391-392 and 399.
IFSTA, *Essentials of Fire Fighting*, 4th Edition, 1st Printing, page 298.
Jones and Bartlett, *Fundamentals of Fire Fighter Skills*, 1st Edition, 1st Printing, pages 340-341.
Answer: B

63. References: NFPA 1001, 5.3.6 and 5.3.6(A)(B)
Delmar, *Firefighter's Handbook*, 2nd Edition, 1st Printing, page 373.
IFSTA, *Essentials of Fire Fighting*, 4th Edition, 1st Printing, page 284.
Jones and Bartlett, *Fundamentals of Fire Fighter Skills*, 1st Edition, 1st Printing, page 333.
Answer: D

64. References: NFPA 1001, 5.3.6 and 5.3.6(A)
Delmar, *Firefighter's Handbook*, 2nd Edition, 1st Printing, page 377.
IFSTA, *Essentials of Fire Fighting*, 4th Edition, 1st Printing, page 283.
Jones and Bartlett, *Fundamentals of Fire Fighter Skills*, 1st Edition, 1st Printing, page 333.
Answer: B

65. References: NFPA 1001, 5.3.6 and 5.3.6(A)
Delmar, *Firefighter's Handbook*, 2nd Edition, 1st Printing, pages 377-378.
IFSTA, *Essentials of Fire Fighting*, 4th Edition, 1st Printing, page 286.
Jones and Bartlett, *Fundamentals of Fire Fighter Skills*, 1st Edition, 1st Printing, page 333.
Answer: A

APPENDIX A. EXAMINATION I-1 ANSWER KEY

66. References: NFPA 1001, 5.3.6 and 5.3.6(A)
Delmar, *Firefighter's Handbook*, 2nd Edition, 1st Printing, page 371.
IFSTA, *Essentials of Fire Fighting*, 4th Edition, 1st Printing, page 281.
Jones and Bartlett, *Fundamentals of Fire Fighter Skills*, 1st Edition, 1st Printing, page 331.
Answer: B

67. References: NFPA 1001, 5.3.6 and 5.3.6(A)
Delmar, *Firefighter's Handbook*, 2nd Edition, 1st Printing, page 373.
IFSTA, *Essentials of Fire Fighting*, 4th Edition, 1st Printing, page 284.
Jones and Bartlett, *Fundamentals of Fire Fighter Skills*, 1st Edition, 1st Printing, page 332.
Answer: A

68. Reference: NFPA 1001, 5.3.7 and 5.3.7(A)(B)
Delmar, *Firefighter's Handbook*, 2nd Edition, 1st Printing, pages 623-624.
IFSTA, *Essentials of Fire Fighting*, 4th Edition, 1st Printing, page 548.
Jones and Bartlett, *Fundamentals of Fire Fighter Skills*, 1st Edition, 1st Printing, page 637.
Answer: C

69. Reference: NFPA 1001, 5.3.7 and 5.3.7(A)(B)
Delmar, *Firefighter's Handbook*, 2nd Edition, 1st Printing, page 624.
IFSTA, *Essentials of Fire Fighting*, 4th Edition, 1st Printing, page 548.
Jones and Bartlett, *Fundamentals of Fire Fighter Skills*, 1st Edition, 1st Printing, pages 637-638.
Answer: D

70. Reference: NFPA 1001, 5.3.7 and 5.3.7(A)(B)
Delmar, *Firefighter's Handbook*, 2nd Edition, 1st Printing, page 513.
IFSTA, *Essentials of Fire Fighting*, 4th Edition, 1st Printing, page 548.
Jones and Bartlett, *Fundamentals of Fire Fighter Skills*, 1st Edition, 1st Printing, page 638.
Answer: A

71. Reference: NFPA 1001, 5.3.7 and 5.3.7(A)(B)
Delmar, *Firefighter's Handbook*, 2nd Edition, 1st Printing, page 513.
IFSTA, *Essentials of Fire Fighting*, 4th Edition, 1st Printing, page 548.
Jones and Bartlett, *Fundamentals of Fire Fighter Skills*, 1st Edition, 1st Printing, page 638.
Answer: A

72. Reference: NFPA 1001, 5.3.8 and 5.3.8(A)(B)
Delmar, *Firefighter's Handbook*, 2nd Edition, 1st Printing, page 100.
IFSTA, *Essentials of Fire Fighting*, 4th Edition, 1st Printing, page 58.
Jones and Bartlett, *Fundamentals of Fire Fighter Skills*, 1st Edition, 1st Printing, page 134.
Answer: A

73. Reference: NFPA 1001, 5.3.8 and 5.3.8(A)
Delmar, *Firefighter's Handbook*, 2nd Edition, 1st Printing, page 203.
IFSTA, *Essentials of Fire Fighting*, 4th Edition, 1st Printing, page 487.
Jones and Bartlett, *Fundamentals of Fire Fighter Skills*, 1st Edition, 1st Printing, page 134.
Answer: C

74. Reference: NFPA 1001, 5.3.8, 5.3.8(A), 5.3.10, and 5.3.10(A)
Delmar, *Firefighter's Handbook*, 2nd Edition, 1st Printing, page 228.
IFSTA, *Essentials of Fire Fighting*, 4th Edition, 1st Printing, page 407.
Jones and Bartlett, *Fundamentals of Fire Fighter Skills*, 1st Edition, 1st Printing, page 481.
Answer: C

75. Reference: NFPA 1001, 5.3.9 and 5.3.9(A)
Delmar, *Firefighter's Handbook*, 2nd Edition, 1st Printing, pages 461 and 506.
IFSTA, *Essentials of Fire Fighting*, 4th Edition, 1st Printing, page 175.
Jones and Bartlett, *Fundamentals of Fire Fighter Skills*, 1st Edition, 1st Printing, pages 374 and 404.
Answer: A

76. Reference: NFPA 1001, 5.3.9 and 5.3.9(A)(B)
Delmar, *Firefighter's Handbook*, 2nd Edition, 1st Printing, page 465.
IFSTA, *Essentials of Fire Fighting*, 4th Edition, 1st Printing, page 177.
Jones and Bartlett, *Fundamentals of Fire Fighter Skills*, 1st Edition, 1st Printing, page 378.
Answer: C

77. Reference: NFPA 1001, 5.3.9 and 5.3.9(A)(B)
Delmar, *Firefighter's Handbook*, 1st Edition, 1st Printing, page 465.
IFSTA, *Essentials of Fire Fighting*, 4th Edition, 1st Printing, pages 176-177.
Jones and Bartlett, *Fundamentals of Fire Fighter Skills*, 1st Edition, 1st Printing, page 378.
Answer: A

78. Reference: NFPA 1001, 5.3.10 and 5.3.10(A)
Delmar, *Firefighter's Handbook*, 2nd Edition, 1st Printing, page 364.
IFSTA, *Essentials of Fire Fighting*, 4th Edition, 1st Printing, pages 73-74.
Jones and Bartlett, *Fundamentals of Fire Fighter Skills*, 1st Edition, 1st Printing, pages 383-384 and 628.
Answer: D

79. Reference: NFPA 1001, 5.3.10, 5.3.10(A), 5.3.1 and 5.3.1(A)
Delmar, *Firefighter's Handbook*, 2nd Edition, 1st Printing, page 145.
IFSTA, *Essentials of Fire Fighting*, 4th Edition, 1st Printing, page 87.
Jones and Bartlett, *Fundamentals of Fire Fighter Skills*, 1st Edition, 1st Printing, page 129.
Answer: A

80. Reference: NFPA 1001, 5.3.10 and 5.3.10(A)(B)
Delmar, *Firefighter's Handbook*, 2nd Edition, 1st Printing, page 290.
IFSTA, *Essentials of Fire Fighting*, 4th Edition, 1st Printing, page 525.
Jones and Bartlett, *Fundamentals of Fire Fighter Skills*, 1st Edition, 1st Printing, page 628.
Answer: B

81. Reference: NFPA 1001, 5.3.10 and 5.3.10(A)(B)
Delmar, *Firefighter's Handbook*, 2nd Edition, 1st Printing, page 227.
IFSTA, *Essentials of Fire Fighting*, 4th Edition, 1st Printing, page 410.
Jones and Bartlett, *Fundamentals of Fire Fighter Skills*, 1st Edition, 1st Printing, page 482.
Answer: D

82. Reference: NFPA 1001, 5.3.10 and 5.3.10(A)(B)
Delmar, *Firefighter's Handbook*, 2nd Edition, 1st Printing, page 283.
IFSTA, *Essentials of Fire Fighting*, 4th Edition, 1st Printing, page 493.
Jones and Bartlett, *Fundamentals of Fire Fighter Skills*, 1st Edition, 1st Printing, page 520.
Answer: A

83. Reference: NFPA 1001, 5.3.10 and 5.3.10(A)(B)
Delmar, *Firefighter's Handbook*, 2nd Edition, 1st Printing, page 289.
IFSTA, *Essentials of Fire Fighting*, 4th Edition, 1st Printing, page 527.
Jones and Bartlett, *Fundamentals of Fire Fighter Skills*, 1st Edition, 1st Printing, page 627.
Answer: D

84. Reference: NFPA 1001, 5.3.10 and 5.3.10(A)(B)
Delmar, *Firefighter's Handbook*, 1st Edition, 1st Printing, page 603.
IFSTA, *Essentials of Fire Fighting*, 4th Edition, 1st Printing, page 525.
Jones and Bartlett, *Fundamentals of Fire Fighter Skills*, 1st Edition, 1st Printing, page 627.
Answer: B

85. Reference: NFPA 1001, 5.3.10, 5.3.10(A)(B), 5.5.4, and 5.5.4(A)(B)
Delmar, *Firefighter's Handbook*, 2nd Edition, 1st Printing, page 231.
IFSTA, *Essentials of Fire Fighting*, 4th Edition, 1st Printing, pages 415 and 445.
Jones and Bartlett, *Fundamentals of Fire Fighter Skills*, 1st Edition, 1st Printing, pages 484 and 488.
Answer: C

86. Reference: NFPA 1001, 5.3.10 and 5.3.10(A)(B)
Delmar, *Firefighter's Handbook*, 2nd Edition, 1st Printing, page 253.
IFSTA, *Essentials of Fire Fighting*, 4th Edition, 1st Printing, pages 430 and 522.
Jones and Bartlett, *Fundamentals of Fire Fighter Skills*, 1st Edition, 1st Printing, page 514.
Answer: A

87. Reference: NFPA 1001, 5.3.10 and 5.3.10(A)(B)
Delmar, *Firefighter's Handbook*, 2nd Edition, 1st Printing, pages 257 and 618.
IFSTA, *Essentials of Fire Fighting*, 4th Edition, 1st Printing, pages 431 and 545.
Jones and Bartlett, *Fundamentals of Fire Fighter Skills*, 1st Edition, 1st Printing, page 518.
Answer: A

88. Reference: NFPA 1001, 5.3.10, 5.3.10(A)(B), 5.3.8 and 5.3.8(A)(B)
Delmar, *Firefighter's Handbook*, 1st Edition, 1st Printing, pages 292 and 305.
IFSTA, *Essentials of Fire Fighting*, 4th Edition, 1st Printing, page 490.
Jones and Bartlett, *Fundamentals of Fire Fighter Skills*, 1st Edition, 1st Printing, page 466.
Answer: B

89. Reference: NFPA 1001, 5.3.10 and 5.3.10(A)(B)
Delmar, *Firefighter's Handbook*, 1st Edition, 1st Printing, page 217.
IFSTA, *Essentials of Fire Fighting*, 4th Edition, 1st Printing, page 491.
Jones and Bartlett, *Fundamentals of Fire Fighter Skills*, 1st Edition, 1st Printing, pages 467 and 533.
Answer: B

90. Reference: NFPA 1001, 5.3.10 and 5.3.10(A)(B)
Delmar, *Firefighter's Handbook*, 2nd Edition, 1st Printing, page 216.
IFSTA, *Essentials of Fire Fighting*, 4th Edition, 1st Printing, page 491.
Jones and Bartlett, *Fundamentals of Fire Fighter Skills*, 1st Edition, 1st Printing, page 520.
Answer: C

91. Reference: NFPA 1001, 5.3.10, 5.3.10(A)(B), 5.3.8 and 5.3.8(A)(B)
Delmar, *Firefighter's Handbook*, 2nd Edition, 1st Printing, pages 283-284.
IFSTA, *Essentials of Fire Fighting*, 4th Edition, 1st Printing, page 496.
Jones and Bartlett, *Fundamentals of Fire Fighter Skills*, 1st Edition, 1st Printing, page 523.
Answer: C

92. Reference: NFPA 1001, 5.3.10, 5.3.10(A), 5.3.8 and 5.3.8(A)(B)
Delmar, *Firefighter's Handbook*, 2nd Edition, 1st Printing, pages 271-274, 288.
IFSTA, *Essentials of Fire Fighting*, 4th Edition, 1st Printing, page 492.
Jones and Bartlett, *Fundamentals of Fire Fighter Skills*, 1st Edition, 1st Printing, page 630.
Answer: D

93. Reference: NFPA 1001, 5.3.10 and 5.3.10(A)(B)
Delmar, *Firefighter's Handbook*, 2nd Edition, 1st Printing, page 284.
IFSTA, *Essentials of Fire Fighting*, 4th Edition, 1st Printing, page 527.
Jones and Bartlett, *Fundamentals of Fire Fighter Skills*, 1st Edition, 1st Printing, page 627.
Answer: D

APPENDIX A, EXAMINATION I-1 ANSWER KEY

94. Reference: NFPA 1001, 5.3.11, 5.3.11(A), 5.3.12, and 5.3.12(A)
Delmar, *Firefighter's Handbook*, 2nd Edition, 1st Printing, page 551.
IFSTA, *Essentials of Fire Fighting*, 4th Edition, 1st Printing, page 346.
Jones and Bartlett, *Fundamentals of Fire Fighter Skills*, 1st Edition, 1st Printing, pages 408, 410, and 414.
Answer: D

95. Reference: NFPA 1001, 5.3.11, 5.3.11(A) 5.3.12, and 5.3.12(A)
Delmar, *Firefighter's Handbook*, 2nd Edition, 1st Printing, page 554.
IFSTA, *Essentials of Fire Fighting*, 4th Edition, 1st Printing, page 364.
Jones and Bartlett, *Fundamentals of Fire Fighter Skills*, 1st Edition, 1st Printing, pages 411 and 415.
Answer: A

96. Reference: NFPA 1001, 5.3.11, 5.3.11(A)(B), 5.3.12, and 5.3.12(A)
Delmar, *Firefighter's Handbook*, 2nd Edition, 1st Printing, page 573.
IFSTA, *Essentials of Fire Fighting*, 4th Edition, 1st Printing, pages 352 and 361-362.
Jones and Bartlett, *Fundamentals of Fire Fighter Skills*, 1st Edition, 1st Printing, page 437.
Answer: B

97. Reference: NFPA 1001, 5.3.11, 5.3.11(A), 5.3.12, 5.3.12(A), 5.3.10 and 5.3.10(A)(B)
Delmar, *Firefighter's Handbook*, 2nd Edition, 1st Printing, page 550.
IFSTA, *Essentials of Fire Fighting*, 4th Edition, 1st Printing, page 345.
Jones and Bartlett, *Fundamentals of Fire Fighter Skills*, 1st Edition, 1st Printing, pages 408 and 410.
Answer: D

98. Reference: NFPA 1001, 5.3.11 and 5.3.11(A)(B)
Delmar, *Firefighter's Handbook*, 2nd Edition, 1st Printing, page 563.
IFSTA, *Essentials of Fire Fighting*, 4th Edition, 1st Printing, page 366.
Jones and Bartlett, *Fundamentals of Fire Fighter Skills*, 1st Edition, 1st Printing, page 411.
Answer: B

99. Reference: NFPA 1001, 5.3.11 and 5.3.11(A)(B)
Delmar, *Firefighter's Handbook*, 2nd Edition, 1st Printing, pages 563-565.
IFSTA, *Essentials of Fire Fighting*, 4th Edition, 1st Printing, pages 369-370.
Jones and Bartlett, *Fundamentals of Fire Fighter Skills*, 1st Edition, 1st Printing, page 411.
Answer: C

100. Reference: NFPA 1001, 5.3.11, 5.3.11(A), 5.3.12 and 5.3.12(A)
Delmar, *Firefighter's Handbook*, 2nd Edition, 1st Printing, pages 95 and 562.
IFSTA, *Essentials of Fire Fighting*, 4th Edition, 1st Printing, page 37.
Jones and Bartlett, *Fundamentals of Fire Fighter Skills*, 1st Edition, 1st Printing, pages 130-131, 136, 408 and 409.
Answer: D

101. Reference: NFPA 1001, 5.3.11, 5.3.11(A)(B), 5.3.12, 5.3.12(A)(B), 5.3.10 and 5.3.10(A).
Delmar, *Firefighter's Handbook*, 2nd Edition, 1st Printing, page 553.
IFSTA, *Essentials of Fire Fighting*, 4th Edition, 1st Printing, page 354.
Jones and Bartlett, *Fundamentals of Fire Fighter Skills*, 1st Edition, 1st Printing, pages 410-411.
Answer: D

102. Reference: NFPA 1001, 5.3.11, 5.3.11(A), 5.3.12, 5.3.12(A), 5.3.13 and 5.3.13(A)
Delmar, *Firefighter's Handbook*, 2nd Edition, 1st Printing, page 76.
IFSTA, *Essentials of Fire Fighting*, 4th Edition, 1st Printing, page 40.
Jones and Bartlett, *Fundamentals of Fire Fighter Skills*, 1st Edition, 1st Printing, pages 124-125.
Answer: B

103. Reference: NFPA 1001, 5.3.12, 5.3.12(A)(B), 5.3.10 and 5.3.10(A)(B)
Delmar, *Firefighter's Handbook*, 2nd Edition, 1st Printing, page 579.
IFSTA, *Essentials of Fire Fighting*, 4th Edition, 1st Printing, page 75.
Jones and Bartlett, *Fundamentals of Fire Fighter Skills*, 1st Edition, 1st Printing, page 427.
Answer: C

104. Reference: NFPA 1001, 5.3.12, 5.3.12(A)(B), 5.3.10 and 5.3.10(A)(B)
Delmar, *Firefighter's Handbook*, 2nd Edition, 1st Printing, page 581.
IFSTA, *Essentials of Fire Fighting*, 4th Edition, 1st Printing, page 363.
Jones and Bartlett, *Fundamentals of Fire Fighter Skills*, 1st Edition, 1st Printing, page 424.
Answer: D

105. Reference: NFPA 1001, 5.3.12 and 5.3.12(A)(B)
Delmar, *Firefighter's Handbook*, 2nd Edition, 1st Printing, page 579.
IFSTA, *Essentials of Fire Fighting*, 4th Edition, 1st Printing, page 361.
Jones and Bartlett, *Fundamentals of Fire Fighter Skills*, 1st Edition, 1st Printing, page 435.
Answer: C

106. Reference: NFPA 1001, 5.3.12 and 5.3.12(A)(B)
Delmar, *Firefighter's Handbook*, 2nd Edition, 1st Printing, page 571.
IFSTA, *Essentials of Fire Fighting*, 4th Edition, 1st Printing, pages 356 and 372.
Jones and Bartlett, *Fundamentals of Fire Fighter Skills*, 1st Edition, 1st Printing, pages 423-425.
Answer: D

107. Reference: NFPA 1001, 5.3.12 and 5.3.12(A)
Delmar, *Firefighter's Handbook*, 2nd Edition, 1st Printing, page 520.
IFSTA, *Essentials of Fire Fighting*, 4th Edition, 1st Printing, page 357.
Jones and Bartlett, *Fundamentals of Fire Fighter Skills*, 1st Edition, 1st Printing, page 431.
Answer: B

108. Reference: NFPA 1001, 5.3.12 and 5.3.12(A)(B)
Delmar, *Firefighter's Handbook*, 2nd Edition, 1st Printing, page 571.
IFSTA, *Essentials of Fire Fighting*, 4th Edition, 1st Printing, pages 356 and 372.
Jones and Bartlett, *Fundamentals of Fire Fighter Skills*, 1st Edition, 1st Printing, pages 423-425.
Answer: C

109. Reference: NFPA 1001, 5.3.12 and 5.3.12(A)
Delmar, *Firefighter's Handbook*, 2nd Edition, 1st Printing, page 99.
IFSTA, *Essentials of Fire Fighting*, 4th Edition, 1st Printing, page 347.
Jones and Bartlett, *Fundamentals of Fire Fighter Skills*, 1st Edition, 1st Printing, pages 408-409.
Answer: B

110. Reference: NFPA 1001, 5.3.13 and 5.3.13(A)(B)
Delmar, *Firefighter's Handbook*, 2nd Edition, 1st Printing, page 639.
IFSTA, *Essentials of Fire Fighting*, 4th Edition, 1st Printing, page 579.
Jones and Bartlett, *Fundamentals of Fire Fighter Skills*, 1st Edition, 1st Printing, pages 555 and 574.
Answer: D

111. Reference: NFPA 1001, 5.3.13 and 5.3.13(A)(B)
Delmar, *Firefighter's Handbook*, 2nd Edition, 1st Printing, page 649.
IFSTA, *Essentials of Fire Fighting*, 4th Edition, 1st Printing, pages 596-597.
Jones and Bartlett, *Fundamentals of Fire Fighter Skills*, 1st Edition, 1st Printing, page 574.
Answer: D

112. Reference: NFPA 1001, 5.3.13 and 5.3.13(A)(B)
Delmar, *Firefighter's Handbook*, 2nd Edition, 1st Printing, page 649.
IFSTA, *Essentials of Fire Fighting*, 4th Edition, 1st Printing, page 597.
Jones and Bartlett, *Fundamentals of Fire Fighter Skills*, 1st Edition, 1st Printing, page 576.
Answer: D

113. Reference: NFPA 1001, 5.3.13 and 5.3.13(A)(B)
Delmar, *Firefighter's Handbook*, 2nd Edition, 1st Printing, pages 518 and 649.
IFSTA, *Essentials of Fire Fighting*, 4th Edition, 1st Printing, pages 596-598.
Jones and Bartlett, *Fundamentals of Fire Fighter Skills*, 1st Edition, 1st Printing, page 577.
Answer: B

114. Reference: NFPA 1001, 5.3.14 and 5.3.14(A)
Delmar, *Firefighter's Handbook*, 2nd Edition, 1st Printing, page 646.
IFSTA, *Essentials of Fire Fighting*, 4th Edition, 1st Printing, page 595.
Jones and Bartlett, *Fundamentals of Fire Fighter Skills*, 1st Edition, 1st Printing, page 569.
Answer: B

115. Reference: NFPA 1001, 5.3.14 and 5.3.14(A)(B)
Delmar, *Firefighter's Handbook*, 2nd Edition, 1st Printing, pages 645-646.
IFSTA, *Essentials of Fire Fighting*, 4th Edition, 1st Printing, page 595.
Jones and Bartlett, *Fundamentals of Fire Fighter Skills*, 1st Edition, 1st Printing, page 569.
Answer: C

116. Reference: NFPA 1001, 5.3.14 and 5.3.14(A)(B)
Delmar, *Firefighter's Handbook*, 2nd Edition, 1st Printing, pages 644-645.
IFSTA, *Essentials of Fire Fighting*, 4th Edition, 1st Printing, pages 594-595.
Jones and Bartlett, *Fundamentals of Fire Fighter Skills*, 1st Edition, 1st Printing, page 560.
Answer: B

117. Reference: NFPA 1001, 5.3.14 and 5.3.14(A)
Delmar, *Firefighter's Handbook*, 2nd Edition, 1st Printing, page 633.
IFSTA, *Essentials of Fire Fighting*, 4th Edition, 1st Printing, page 587.
Jones and Bartlett, *Fundamentals of Fire Fighter Skills*, 1st Edition, 1st Printing, page 554.
Answer: C

118. Reference: NFPA 1001, 5.3.14 and 5.3.14(A)(B)
Delmar, *Firefighter's Handbook*, 2nd Edition, 1st Printing, pages 645-646.
IFSTA, *Essentials of Fire Fighting*, 4th Edition, 1st Printing, page 617.
Jones and Bartlett, *Fundamentals of Fire Fighter Skills*, 1st Edition, 1st Printing, pages 571 and 573.
Answer: C

119. Reference: NFPA 1001, 5.3.14 and 5.3.14(A)(B)
Delmar, *Firefighter's Handbook*, 2nd Edition, 1st Printing, page 633.
IFSTA, *Essentials of Fire Fighting*, 4th Edition, 1st Printing, page 587.
Jones and Bartlett, *Fundamentals of Fire Fighter Skills*, 1st Edition, 1st Printing, page 554.
Answer: C

120. Reference: NFPA 1001, 5.3.14 and 5.3.14(A)(B)
Delmar, *Firefighter's Handbook*, 2nd Edition, 1st Printing, pages 634-635.
IFSTA, *Essentials of Fire Fighting*, 4th Edition, 1st Printing, pages 592-593.
Jones and Bartlett, *Fundamentals of Fire Fighter Skills*, 1st Edition, 1st Printing, page 565.
Answer: D

121. Reference: NFPA 1001, 5.3.15 and 5.3.15(A)(B)
Delmar, *Firefighter's Handbook*, 2nd Edition, 1st Printing, page 208.
IFSTA, *Essentials of Fire Fighting*, 4th Edition, 1st Printing, page 464.
Jones and Bartlett, *Fundamentals of Fire Fighter Skills*, 1st Edition, 1st Printing, page 451.
Answer: B

122. Reference: NFPA 1001, 5.3.15 and 5.3.15(A)(B)
Delmar, *Firefighter's Handbook*, 2nd Edition, 1st Printing, page 270.
IFSTA, *Essentials of Fire Fighting*, 4th Edition, 1st Printing, page 423.
Jones and Bartlett, *Fundamentals of Fire Fighter Skills*, 1st Edition, 1st Printing, pages 489-490.
Answer: B

123. Reference: NFPA 1001, 5.3.15 and 5.3.15(A)(B)
Delmar, *Firefighter's Handbook*, 2nd Edition, 1st Printing, page 270.
IFSTA, *Essentials of Fire Fighting*, 4th Edition, 1st Printing, page 425.
Jones and Bartlett, *Fundamentals of Fire Fighter Skills*, 1st Edition, 1st Printing, pages 490-491.
Answer: B

124. Reference: NFPA 1001, 5.3.15 and 5.3.15(A)
Delmar, *Firefighter's Handbook*, 2nd Edition, 1st Printing, page 223.
IFSTA, *Essentials of Fire Fighting*, 4th Edition, 1st Printing, page 397.
Jones and Bartlett, *Fundamentals of Fire Fighter Skills*, 1st Edition, 1st Printing, page 477.
Answer: A

125. Reference: NFPA 1001, 5.3.15 and 5.3.15(A)
Delmar, *Firefighter's Handbook*, 2nd Edition, 1st Printing, page 270.
IFSTA, *Essentials of Fire Fighting*, 4th Edition, 1st Printing, page 423.
Jones and Bartlett, *Fundamentals of Fire Fighter Skills*, 1st Edition, 1st Printing, page 489-490.
Answer: D

126. Reference: NFPA 1001, 5.3.15 and 5.3.15(A)
Delmar, *Firefighter's Handbook*, 2nd Edition, 1st Printing, page 223.
IFSTA, *Essentials of Fire Fighting*, 4th Edition, 1st Printing, page 415.
Jones and Bartlett, *Fundamentals of Fire Fighter Skills*, 1st Edition, 1st Printing, page 492.
Answer: C

127. Reference: NFPA 1001, 5.3.16 and 5.3.16(A)(B)
Delmar, *Firefighter's Handbook*, 2nd Edition, 1st Printing, page 188.
IFSTA, *Essentials of Fire Fighting*, 4th Edition, 1st Printing, pages 126 and 128.
Jones and Bartlett, *Fundamentals of Fire Fighter Skills*, 1st Edition, 1st Printing, page 187.
Answer: C

128. Reference: NFPA 1001, 5.3.16 and 5.3.16(A)
Delmar, *Firefighter's Handbook*, 2nd Edition, 1st Printing, page 185.
IFSTA, *Essentials of Fire Fighting*, 4th Edition, 1st Printing, page 133.
Jones and Bartlett, *Fundamentals of Fire Fighter Skills*, 1st Edition, 1st Printing, page 178.
Answer: C

129. Reference: NFPA 1001, 5.3.16 and 5.3.16(A)(B)
Delmar, *Firefighter's Handbook*, 2nd Edition, 1st Printing, page 194.
IFSTA, *Essentials of Fire Fighting*, 4th Edition, 1st Printing, page 134.
Jones and Bartlett, *Fundamentals of Fire Fighter Skills*, 1st Edition, 1st Printing, page 180.
Answer: B

130. Reference: NFPA 1001, 5.3.16 and 5.3.16(A)(B)
Delmar, *Firefighter's Handbook*, 2nd Edition, 1st Printing, page 188.
IFSTA, *Essentials of Fire Fighting*, 4th Edition, 1st Printing, page 136.
Jones and Bartlett, *Fundamentals of Fire Fighter Skills*, 1st Edition, 1st Printing, page 188.
Answer: B

131. Reference: NFPA 1001, 5.3.16 and 5.3.16(A)
Delmar, *Firefighter's Handbook*, 2nd Edition, 1st Printing, page 189.
IFSTA, *Essentials of Fire Fighting*, 4th Edition, 1st Printing, pages 133 and 135.
Jones and Bartlett, *Fundamentals of Fire Fighter Skills*, 1st Edition, 1st Printing, page 181.
Answer: C

132. Reference: NFPA 1001, 5.3.16 and 5.3.16(A)
Delmar, *Firefighter's Handbook*, 2nd Edition, 1st Printing, page 189.
IFSTA, *Essentials of Fire Fighting*, 4th Edition, 1st Printing, pages 133-135.
Jones and Bartlett, *Fundamentals of Fire Fighter Skills*, 1st Edition, 1st Printing, page 181.
Answer: A

133. Reference: NFPA 1001, 5.3.16 and 5.3.16(A)
Delmar, *Firefighter's Handbook*, 2nd Edition, 1st Printing, page 186.
IFSTA, *Essentials of Fire Fighting*, 4th Edition, 1st Printing, pages 126 and 132.
Jones and Bartlett, *Fundamentals of Fire Fighter Skills*, 1st Edition, 1st Printing, pages 188 and 193.
Answer: C

134. Reference: NFPA 1001, 5.3.17 and 5.3.17(A)(B)
IFSTA, *Essentials of Fire Fighting*, 4th Edition, 1st Printing, page 187.
Jones and Bartlett, *Fundamentals of Fire Fighter Skills*, 1st Printing, 1st Edition, page 581.
Answer: C

135. Reference: NFPA 1001, 5.3.18 and 5.3.18(A)(B)
Delmar, *Firefighter's Handbook*, 2nd Edition, 1st Printing, page 498.
IFSTA, *Essentials of Fire Fighting*, 4th Edition, 1st Printing, page 210.
Jones and Bartlett, *Fundamentals of Fire Fighter Skills*, 1st Edition, 1st Printing, page 642.
Answer: B

136. Reference: NFPA 1001, 5.3.18 and 5.3.18(A)(B)
Delmar, *Firefighter's Handbook*, 2nd Edition, 1st Printing, page 498.
IFSTA, *Essentials of Fire Fighting*, 4th Edition, 1st Printing, pages 210 and 540.
Jones and Bartlett, *Fundamentals of Fire Fighter Skills*, 1st Edition, 1st Printing, page 28.
Answer: A

137. Reference: NFPA 1001, 5.3.19 and 5.3.19(A)(B)
Delmar, *Firefighter's Handbook*, 2nd Edition, 1st Printing, pages 598-599.
IFSTA, *Essentials of Fire Fighting*, 4th Edition, 1st Printing, page 554.
Jones and Bartlett, *Fundamentals of Fire Fighter Skills*, 1st Edition, 1st Printing, page 611.
Answer: C

138. Reference: NFPA 1001, 5.3.19 and 5.3.19(A)
Delmar, *Firefighter's Handbook*, 2nd Edition, 1st Printing, page 621.
IFSTA, *Essentials of Fire Fighting*, 4th Edition, 1st Printing, page 556.
Jones and Bartlett, *Fundamentals of Fire Fighter Skills*, 1st Edition, 1st Printing, page 613.
Answer: B

139. Reference: NFPA 1001, 5.3.19 and 5.3.19(A)
Delmar, *Firefighter's Handbook*, 2nd Edition, 1st Printing, pages 595-599.
IFSTA, *Essentials of Fire Fighting*, 4th Edition, 1st Printing, page 554.
Jones and Bartlett, *Fundamentals of Fire Fighter Skills*, 1st Edition, 1st Printing, pages 609-611.
Answer: C

140. Reference: NFPA 1001, 5.5.1 and 5.5.1(A)
Delmar, *Firefighter's Handbook*, 2nd Edition, 1st Printing, pages 685 and 687.
IFSTA, *Essentials of Fire Fighting*, 4th Edition, 1st Printing, page 659.
Jones and Bartlett, *Fundamentals of Fire Fighter Skills*, 1st Edition, 1st Printing, page 650.
Answer: C

141. Reference: NFPA 1001, 5.5.1 and 5.5.1(A)(B)
Delmar, *Firefighter's Handbook*, 2nd Edition, 1st Printing, page 660.
IFSTA, *Essentials of Fire Fighting*, 4th Edition, 1st Printing, page 662.
Jones and Bartlett, *Fundamentals of Fire Fighter Skills*, 1st Edition, 1st Printing, pages 650-651.
Answer: C

142. Reference: NFPA 1001, 5.5.1 and 5.5.1(A)
Delmar, *Firefighter's Handbook*, 2nd Edition, 1st Printing, page 660.
IFSTA, *Essentials of Fire Fighting*, 4th Edition, 1st Printing, page 662.
Jones and Bartlett, *Fundamentals of Fire Fighter Skills*, 1st Edition, 1st Printing, pages 648 and 912.
Answer: D

143. Reference: NFPA 1001, 5.5.2 and 5.5.2(A)(B)
Delmar, *Firefighter's Handbook*, 2nd Edition, 1st Printing, page 32.
IFSTA, *Essentials of Fire Fighting*, 4th Edition, 1st Printing, pages 655-656.
Jones and Bartlett, *Fundamentals of Fire Fighter Skills*, 1st Edition, 1st Printing, page 911.
Answer: B

144. Reference: NFPA 1001, 5.5.2 and 5.5.2(A)
Delmar, *Firefighter's Handbook*, 2nd Edition, 1st Printing, page 659.
IFSTA, *Essentials of Fire Fighting*, 4th Edition, 1st Printing, page 656.
Jones and Bartlett, *Fundamentals of Fire Fighter Skills*, 1st Edition, 1st Printing, page 910.
Answer: A

145. Reference: NFPA 1001, 5.5.2 and 5.5.2(A)(B)
Delmar, *Firefighter's Handbook*, 2nd Edition, 1st Printing, page 640.
IFSTA, *Essentials of Fire Fighting*, 4th Edition, 1st Printing, page 672.
Jones and Bartlett, *Fundamentals of Fire Fighter Skills*, 1st Edition, 1st Printing, page 917.
Answer: B

146. Reference: NFPA 1001, 5.5.3 and 5.5.3(A)(B)
Delmar, *Firefighter's Handbook*, 2nd Edition, 1st Printing, page 443.
IFSTA, *Essentials of Fire Fighting*, 4th Edition, 1st Printing, page 153.
Jones and Bartlett, *Fundamentals of Fire Fighter Skills*, 1st Edition, 1st Printing, page 252.
Answer: C

147. Reference: NFPA 1001, 5.5.3 and 5.5.3(A)(B)
Delmar, *Firefighter's Handbook*, 2nd Edition, 1st Printing, page 177.
IFSTA, *Essentials of Fire Fighting*, 4th Edition, 1st Printing, page 106.
Jones and Bartlett, *Fundamentals of Fire Fighter Skills*, 1st Edition, 1st Printing, pages 62 and 64.
Answer: B

148. Reference: NFPA 1001, 5.5.3 and 5.5.3(A)(B)
Delmar, *Firefighter's Handbook*, 2nd Edition, 1st Printing, page 172.
IFSTA, *Essentials of Fire Fighting*, 4th Edition, 1st Printing, pages 105-106.
Jones and Bartlett, *Fundamentals of Fire Fighter Skills*, 1st Edition, 1st Printing, page 62.
Answer: A

149. Reference: NFPA 1001, 5.5.4 and 5.5.4(A)(B)
Delmar, *Firefighter's Handbook*, 2nd Edition, 1st Printing, page 224.
IFSTA, *Essentials of Fire Fighting*, 4th Edition, 1st Printing, page 401.
Jones and Bartlett, *Fundamentals of Fire Fighter Skills*, 1st Edition, 1st Printing, page 479.
Answer: C

150. Reference: NFPA 1001, 5.5.4 and 5.5.4(A)(B)
Delmar, *Firefighter's Handbook*, 2nd Edition, 1st Printing, page 224.
IFSTA, *Essentials of Fire Fighting*, 4th Edition, 1st Printing, page 401.
Jones and Bartlett, *Fundamentals of Fire Fighter Skills*, 1st Edition, 1st Printing, page 479.
Answer: B

Examination I-2 Answer Key

Directions
Follow these steps carefully for completing the feedback step of Systematic Approach to Exam Preparation (SAEP):

1. After entering your scores, look up the answers for the examination items you missed, as well as those you guessed, even if you guessed correctly. If you are guessing, it means the answer is not perfectly clear. In this process we are committed to making you as knowledgeable as possible.
2. Enter the number of missed and guessed examination items in the blank on the personal progress plotter.
3. Highlight the answer in the reference materials. Once you have highlighted the answer, read the paragraph preceding and the paragraph following the one in which the correct answer is located. Now that you have highlighted the answer, enter the paragraph number and page number next to the guessed or missed examination item on your examination. Count any part of a paragraph at the beginning of the page as one paragraph until you reach the paragraph containing your highlighted answer. This step will help you locate and review your missed and guessed examination items later in the process. This step is __essential__ to learning the material in context and by association. These learning techniques (context/association) are the very backbone of the SAEP approach.
4. Once you have completed the feedback step you may proceed to the next examination.

1. Reference: NFPA 1001, 5.1.1.1
   Delmar, *Firefighter's Handbook*, 2nd Edition, 1st Printing, page 29.
   IFSTA, *Essentials of Fire Fighting*, 4th Edition, 1st Printing, page 8.
   Jones and Bartlett, *Fundamentals of Fire Fighter Skills*, 1st Edition, 1st Printing, pages 9-10.
   Answer: D

2. Reference: NFPA 1001, 5.1.1.1
   Delmar, *Firefighter's Handbook*, 2nd Edition, 1st Printing, page 32.
   IFSTA, *Essentials of Fire Fighting*, 4th Edition, 1st Printing, pages 6 and 7.
   Jones and Bartlett, *Fundamentals of Fire Fighter Skills*, 1st Edition, 1st Printing, page 9.
   Answer: B

3. Reference: NFPA 1001, 5.1.1.1
   Delmar, *Firefighter's Handbook*, 2nd Edition, 1st Printing, page 33.
   IFSTA, *Essentials of Fire Fighting*, 4th Edition, 1st Printing, page 12.
   Jones and Bartlett, *Fundamentals of Fire Fighter Skills*, 1st Edition, 1st Printing, page 7.
   Answer: C

4. Reference: NFPA 1001, 5.1.1.1
Delmar, *Firefighter's Handbook*, 2nd Edition, 1st Printing, pages 34-35.
IFSTA, *Essentials of Fire Fighting*, 4th Edition, 1st Printing, page 12.
Jones and Bartlett, *Fundamentals of Fire Fighter Skills*, 1st Edition, 1st Printing, page 13.
Answer: A

5. Reference: NFPA 1001, 5.1.1.1
Delmar, *Firefighter's Handbook*, 2nd Edition, 1st Printing, page 35.
IFSTA, *Essentials of Fire Fighting*, 4th Edition, 1st Printing, page 12.
Jones and Bartlett, *Fundamentals of Fire Fighter Skills*, 1st Edition, 1st Printing, page 13.
Answer: D

6. Reference: NFPA 1001, 5.1.1.1
Delmar, *Firefighter's Handbook*, 2nd Edition, 1st Printing, page 39.
IFSTA, *Essentials of Fire Fighting*, 4th Edition, 1st Printing, page 15.
Jones and Bartlett, *Fundamentals of Fire Fighter Skills*, 1st Edition, 1st Printing, pages 108-109.
Answer: B

7. Reference: NFPA 1001, 5.1.1.1
Delmar, *Firefighter's Handbook*, 2nd Edition, 1st Printing, page 109.
IFSTA, *Essentials of Fire Fighting*, 4th Edition, 1st Printing, page 25.
Jones and Bartlett, *Fundamentals of Fire Fighter Skills*, 1st Edition, 1st Printing, page 24.
Answer: C

8. Reference: NFPA 1001, 5.1.1.1
Delmar, *Firefighter's Handbook*, 2nd Edition, 1st Printing, page 39.
IFSTA, *Essentials of Fire Fighting*, 4th Edition, 1st Printing, page 14.
Jones and Bartlett, *Fundamentals of Fire Fighter Skills*, 1st Edition, 1st Printing, page 106.
Answer: D

9. Reference: NFPA 1001, 5.1.1.1
Delmar, *Firefighter's Handbook*, 2nd Edition, 1st Printing, pages 39-40.
IFSTA, *Essentials of Fire Fighting*, 4th Edition, 1st Printing, page 15.
Jones and Bartlett, *Fundamentals of Fire Fighter Skills*, 1st Edition, 1st Printing, page 109.
Answer: A

10. Reference: NFPA 1001, 5.1.1.1, and 5.1.1.2
Delmar, *Firefighter's Handbook*, 2nd Edition, 1st Printing, page 426.
IFSTA, *Essentials of Fire Fighting*, 4th Edition, 1st Printing, page 155.
Jones and Bartlett, *Fundamentals of Fire Fighter Skills*, 1st Edition, 1st Printing, page 255.
Answer: B

11. Reference: NFPA 1001, 5.1.1.1, and 5.1.1.2
Delmar, *Firefighter's Handbook*, 2nd Edition, 1st Printing, page 426.
IFSTA, *Essentials of Fire Fighting*, 4th Edition, 1st Printing, pages 154-155.
Jones and Bartlett, *Fundamentals of Fire Fighter Skills*, 1st Edition, 1st Printing, page 255.
Answer: A

12. Reference: NFPA 1001, 5.1.1.1, and 5.1.1.2
Delmar, *Firefighter's Handbook*, 2nd Edition, 1st Printing, page 428.
IFSTA, *Essentials of Fire Fighting*, 4th Edition, 1st Printing, page 155.
Jones and Bartlett, *Fundamentals of Fire Fighter Skills*, 1st Edition, 1st Printing, page 258.
Answer: B

13. Reference: NFPA 1001, 5.1.1.1, and 5.1.1.2
Delmar, *Firefighter's Handbook*, 2nd Edition, 1st Printing, page 426.
IFSTA, *Essentials of Fire Fighting*, 4th Edition, 1st Printing, page 154.
Jones and Bartlett, *Fundamentals of Fire Fighter Skills*, 1st Edition, 1st Printing, page 255.
Answer: D

14. Reference: NFPA 1001, 5.1.1.1, and 5.1.1.2
Delmar, *Firefighter's Handbook*, 2nd Edition, 1st Printing, page 451.
IFSTA, *Essentials of Fire Fighting*, 4th Edition, 1st Printing, pages 154 and 165.
Jones and Bartlett, *Fundamentals of Fire Fighter Skills*, 1st Edition, 1st Printing, page 268.
Answer: D

15. Reference: NFPA 1001, 5.1.1.1, and 5.1.1.2
IFSTA, *Essentials of Fire Fighting*, 4th Edition, 1st Printing, page 156.
Answer: A

16. Reference: NFPA 1001, 5.1.1.1
Delmar, *Firefighter's Handbook*, 2nd Edition, 1st Printing, page 422.
IFSTA, *Essentials of Fire Fighting*, 4th Edition, 1st printing, page 150.
Jones and Bartlett, *Fundamentals of Fire Fighter Skills*, 1st Edition, 1st Printing, page 246.
Answer: B

17. Reference: NFPA 1001, 5.1.1.1
Delmar, *Firefighter's Handbook*, 2nd Edition, 1st Printing, page 425.
IFSTA, *Essentials of Fire Fighting*, 4th Edition, 1st Printing, page 154.
Jones and Bartlett, *Fundamentals of Fire Fighter Skills*, 1st Edition, 1st Printing, page 254.
Answer: D

18. Reference: NFPA 1001, 5.1.1.1
Delmar, *Firefighter's Handbook*, 2nd Edition, 1st Printing, page 439.
IFSTA, *Essentials of Fire Fighting*, 4th Edition, 1st Printing, pages 147-148.
Jones and Bartlett, *Fundamentals of Fire Fighter Skills*, 1st Edition, 1st Printing, page 252.
Answer: B

19. Reference: NFPA 1001, 5.2.1 and 5.2.1(A)(B)
Delmar, *Firefighter's Handbook*, 2nd Edition, 1st Printing, page 67.
IFSTA, *Essentials of Fire Fighting*, 4th Edition, 1st Printing, page 648.
Jones and Bartlett, *Fundamentals of Fire Fighter Skills*, 1st Edition, 1st Printing, pages 83-85.
Answer: C

20. Reference: NFPA 1001, 5.2.1 and 5.2.1(A)
Delmar, *Firefighter's Handbook*, 2nd Edition, 1st Printing, page 54.
IFSTA, *Essentials of Fire Fighting*, 4th Edition, 1st Printing, page 641.
Jones and Bartlett, *Fundamentals of Fire Fighter Skills*, 1st Edition, 1st Printing, pages 77-78.
Answer: A

21. Reference: NFPA 1001, 5.2.1 and 5.2.1(A)(B)
Delmar, *Firefighter's Handbook*, 2nd Edition, 1st Printing, page 68.
IFSTA, *Essentials of Fire Fighting*, 4th Edition, 1st Printing, page 639.
Jones and Bartlett, *Fundamentals of Fire Fighter Skills*, 1st Edition, 1st Printing, page 75.
Answer: A

22. Reference: NFPA 1001, 5.2.3 and 5.2.3(A)(B)
IFSTA, *Essentials of Fire Fighting*, 4th Edition, 1st Printing, page 649.
Jones and Bartlett, *Fundamentals of Fire Fighter Skills*, 1st Edition, 1st Printing, pages 87-89.
Answer: B

23. Reference: NFPA 1001, 5.3.1, 5.3.1(A)(B), and 5.3.5(A)(B)
Delmar, *Firefighter's Handbook*, 2nd Edition, 1st Printing, pages 151 and 158.
IFSTA, *Essentials of Fire Fighting*, 4th Edition, 1st Printing, page 95.
Jones and Bartlett, *Fundamentals of Fire Fighter Skills*, 1st Edition, 1st Printing, page 45.
Answer: B

24. Reference: NFPA 1001, 5.3.1, 5.3.1(A)(B), 5.3.5, 5.3.5(A)(B), 5.3.9, and 5.3.9(A)
Delmar, *Firefighter's Handbook*, 2nd Edition, 1st Printing, page 157.
IFSTA, *Essentials of Fire Fighting*, 4th Edition, 1st Printing, page 98.
Jones and Bartlett, *Fundamentals of Fire Fighter Skills*, 1st Edition, 1st Printing, page 58.
Answer: C

25. Reference: NFPA 1001, 5.3.1, 5.3.1(A)(B), 5.3.5, and 5.3.5(A)(B)
Delmar, *Firefighter's Handbook*, 2nd Edition, 1st Printing, pages 150-151.
IFSTA, *Essentials of Fire Fighting*, 4th Edition, 1st Printing, pages 93-94.
Jones and Bartlett, *Fundamentals of Fire Fighter Skills*, 1st Edition, 1st Printing, page 47.
Answer: D

26. Reference: NFPA 1001, 5.3.1 and 5.3.1(A)(B)
Delmar, *Firefighter's Handbook*, 2nd Edition, 1st Printing, page 177.
IFSTA, *Essentials of Fire Fighting*, 4th Edition, 1st Printing, page 113.
Jones and Bartlett, *Fundamentals of Fire Fighter Skills*, 1st Edition, 1st Printing, page 64, Fig. 2-32.
Answer: A

27. Reference: NFPA 1001, 5.3.1, 5.3.1(A)(B), 5.3.5, and 5.3.5(A)(B)
Delmar, *Firefighter's Handbook*, 2nd Edition, 1st Printing, page 135.
IFSTA, *Essentials of Fire Fighting*, 4th Edition, 1st Printing, page 99.
Jones and Bartlett, *Fundamentals of Fire Fighter Skills*, 1st Edition, 1st Printing, pages 35 and 67.
Answer: C

28. Reference: NFPA 1001, 5.3.1, 5.3.1(A)(B), 5.3.5, 5.3.5(A)(B), 5.3.9, and 5.3.9(A)(B)
Delmar, *Firefighter's Handbook*, 2nd Edition, 1st Printing, page 157.
IFSTA, *Essentials of Fire Fighting*, 4th Edition, 1st Printing, page 98.
Jones and Bartlett, *Fundamentals of Fire Fighter Skills*, 1st Edition, 1st Printing, page 47.
Answer: B

29. Reference: NFPA 1001, 5.3.1, 5.3.1(A), 5.3.5, and 5.3.5(A)(B)
Delmar, *Firefighter's Handbook*, 2nd Edition, 1st Printing, page 143.
IFSTA, *Essentials of Fire Fighting*, 4th Edition, 1st Printing, pages 88-89.
Jones and Bartlett, *Fundamentals of Fire Fighter Skills*, 1st Edition, 1st Printing, page 43.
Answer: A

30. Reference: NFPA 1001, 5.3.1 and 5.3.1(A)(B)
Delmar, *Firefighter's Handbook*, 2nd Edition, 1st Printing, page 143.
IFSTA, *Essentials of Fire Fighting*, 4th Edition, 1st Printing, page 87.
Jones and Bartlett, *Fundamentals of Fire Fighter Skills*, 1st Edition, 1st Printing, pages 42-44.
Answer: B

31. Reference: NFPA 1001, 5.3.2 and 5.3.2(A)(B)
Delmar, *Firefighter's Handbook*, 2nd Edition, 1st Printing, page 136.
IFSTA, *Essentials of Fire Fighting*, 4th Edition, 1st Printing, page 86.
Jones and Bartlett, *Fundamentals of Fire Fighter Skills*, 1st Edition, 1st Printing, page 40.
Answer: A

32. Reference: NFPA 1001, 5.3.2, 5.3.2(A), 5.3.3 and 5.3.3(A)(B)
Delmar, *Firefighter's Handbook*, 2nd Edition, 1st Printing, page 128.
IFSTA, *Essentials of Fire Fighting*, 4th Edition, 1st Printing, page 83.
Jones and Bartlett, *Fundamentals of Fire Fighter Skills*, 1st Edition, 1st Printing, pages 33-34.
Answer: A

33. Reference: NFPA 1001, 5.3.2 and 5.3.2(A)(B)
Delmar, *Firefighter's Handbook*, 2nd Edition, 1st Printing, page 135.
IFSTA, *Essentials of Fire Fighting*, 4th Edition, 1st Printing, pages 80 and 100.
Jones and Bartlett, *Fundamentals of Fire Fighter Skills*, 1st Edition, 1st Printing, page 37.
Answer: B

34. Reference: NFPA 1001, 5.3.3 and 5.3.3(A)(B)
IFSTA, *Essentials of Fire Fighting*, 4th Edition, 1st Printing, page 29.
Jones and Bartlett, *Fundamentals of Fire Fighter Skills*, 1st Edition, 1st Printing, page 487.
Answer: A

35. Reference: NFPA 5.3.4 and 5.3.4(A)(B)
Delmar, *Firefighter's Handbook*, 2nd Edition, 1st Printing, page 526.
IFSTA, *Essentials of Fire Fighting*, 4th Edition, 1st Printing, page 249.
Jones and Bartlett, *Fundamentals of Fire Fighter Skills*, 1st Edition, 1st Printing, page 311.
Answer: D

36. Reference: NFPA 5.3.4 and 5.3.4(A)(B)
Delmar, *Firefighter's Handbook*, 2nd Edition, 1st Printing, page 519.
IFSTA, *Essentials of Fire Fighting*, 4th Edition, 1st Printing, page 255-256.
Jones and Bartlett, *Fundamentals of Fire Fighter Skills*, 1st Edition, 1st Printing, pages 230 and 303.
Answer: B

37. Reference: NFPA 5.3.4 and 5.3.4(A)
Delmar, *Firefighter's Handbook*, 2nd Edition, 1st Printing, page 522.
IFSTA, *Essentials of Fire Fighting*, 4th Edition, 1st Printing, pages 244-255.
Jones and Bartlett, *Fundamentals of Fire Fighter Skills*, 1st Edition, 1st Printing, page 304.
Answer: A

38. Reference: NFPA 5.3.4 and 5.3.4(A)(B)
Delmar, *Firefighter's Handbook*, 2nd Edition, 1st Printing, page 524, Fig. 17-26.
IFSTA, *Essentials of Fire Fighting*, 4th Edition, 1st Printing, page 243.
Jones and Bartlett, *Fundamentals of Fire Fighter Skills*, 1st Edition, 1st Printing, pages 305-307.
Answer: B

39. Reference: NFPA 5.3.4 and 5.3.4(A)
Delmar, *Firefighter's Handbook*, 2nd Edition, 1st Printing, page 513.
IFSTA, *Essentials of Fire Fighting*, 4th Edition, 1st Printing, page 236.
Jones and Bartlett, *Fundamentals of Fire Fighter Skills*, 1st Edition, 1st Printing, pages 300-301.
Answer: D

40. Reference: NFPA 5.3.4 and 5.3.4(A)(B)
Delmar, *Firefighter's Handbook*, 2nd Edition, 1st Printing, page 542.
IFSTA, *Essentials of Fire Fighting*, 4th Edition, 1st Printing, page 268.
Jones and Bartlett, *Fundamentals of Fire Fighter Skills*, 1st Edition, 1st Printing, page 311.
Answer: D

41. Reference: NFPA 5.3.4 and 5.3.4(A)(B)
Delmar, *Firefighter's Handbook*, 2nd Edition, 1st Printing, page 542.
IFSTA, *Essentials of Fire Fighting*, 4th Edition, 1st Printing, page 268.
Jones and Bartlett, *Fundamentals of Fire Fighter Skills*, 1st Edition, 1st Printing, page 311.
Answer: B

42. Reference: NFPA 5.3.4 and 5.3.4(A)(B)
Delmar, *Firefighter's Handbook*, 2nd Edition, 1st Printing, page 542.
IFSTA, *Essentials of Fire Fighting*, 4th Edition, 1st Printing, pages 261 and 268.
Jones and Bartlett, *Fundamentals of Fire Fighter Skills*, 1st Edition, 1st Printing, page 311.
Answer: D

43. Reference: NFPA 5.3.4 and 5.3.4(A)(B)
Delmar, *Firefighter's Handbook*, 2nd Edition, 1st Printing, page 543.
IFSTA, *Essentials of Fire Fighting*, 4th Edition, 1st Printing, page 276.
Jones and Bartlett, *Fundamentals of Fire Fighter Skills*, 1st Edition, 1st Printing, page 312.
Answer: B

44. Reference: NFPA 5.3.4 and 5.3.4(A)(B)
Delmar, *Firefighter's Handbook*, 2nd Edition, 1st Printing, page 525.
IFSTA, *Essentials of Fire Fighting*, 4th Edition, 1st Printing, page 248.
Jones and Bartlett, *Fundamentals of Fire Fighter Skills*, 1st Edition, 1st Printing, page 311.
Answer: B

45. Reference: NFPA 5.3.4 and 5.3.4(A)(B)
Delmar, *Firefighter's Handbook*, 2nd Edition, 1st Printing, page 544.
IFSTA, *Essentials of Fire Fighting*, 4th Edition, 1st Printing, pages 264-265.
Jones and Bartlett, *Fundamentals of Fire Fighter Skills*, 1st Edition, 1st Printing, page 323.
Answer: A

46. Reference: NFPA 1001, 5.3.5 and 5.3.5(A)(B)
Delmar, *Firefighter's Handbook*, 2nd Edition, 1st Printing, page 736.
IFSTA, *Essentials of Fire Fighting*, 4th Edition, 1st Printing, page 181.
Jones and Bartlett, *Fundamentals of Fire Fighter Skills*, 1st Edition, 1st Printing, page 545.
Answer: D

47. Reference: NFPA 1001, 5.3.5 and 5.3.5(A)
Delmar, *Firefighter's Handbook*, 2nd Edition, 1st Printing, page 725.
IFSTA, *Essentials of Fire Fighting*, 4th Edition, 1st Printing, pages 29-30.
Jones and Bartlett, *Fundamentals of Fire Fighter Skills*, 1st Edition, 1st Printing, pages 540-541.
Answer: A

48. Reference: NFPA 1001, 5.3.5 and 5.3.5(A)
Delmar, *Firefighter's Handbook*, 2nd Edition, 1st Printing, page 725.
IFSTA, *Essentials of Fire Fighting*, 4th Edition, 1st Printing, pages 29-30.
Jones and Bartlett, *Fundamentals of Fire Fighter Skills*, 1st Edition, 1st Printing, pages 285 and 540.
Answer: A

49. References: NFPA 1001, 5.3.6, 5.3.6(A)(B), 5.3.12 and 5.3.12(A)(B)
Delmar, *Firefighter's Handbook*, 2nd Edition, 1st Printing, page 395.
IFSTA, *Essentials of Fire Fighting*, 4th Edition, 1st Printing, page 311.
Jones and Bartlett, *Fundamentals of Fire Fighter Skills*, 1st Edition, 1st Printing, pages 348-349.
Answer: D

50. References: NFPA 1001, 5.3.6, 5.3.6(A)(B), 5.3.11, 5.3.11(A)(B) 5.3.12, and 5.3.12(A)(B)
Delmar, *Firefighter's Handbook*, 2nd Edition, 1st Printing, page 399.
IFSTA, *Essentials of Fire Fighting*, 4th Edition, 1st Printing, pages 292 and 298.
Jones and Bartlett, *Fundamentals of Fire Fighter Skills*, 1st Edition, 1st Printing, pages 340-341.
Answer: B

51. References: NFPA 1001, 5.3.6, 5.3.6(A)(B), 5.3.12 and 5.3.12(A)(B)
Delmar, *Firefighter's Handbook*, 2nd Edition, 1st Printing, page 402.
IFSTA, *Essentials of Fire Fighting*, 4th Edition, 1st Printing, page 300.
Jones and Bartlett, *Fundamentals of Fire Fighter Skills*, 1st Edition, 1st Printing, pages 353-354.
Answer: B

52. References: NFPA 1001, 5.3.6, 5.3.6(A)(B), 5.3.12, and 5.3.12(A)(B)
Delmar, *Firefighter's Handbook*, 2nd Edition, 1st Printing, page 390.
IFSTA, *Essentials of Fire Fighting*, 4th Edition, 1st Printing, page 299.
Jones and Bartlett, *Fundamentals of Fire Fighter Skills*, 1st Edition, 1st Printing, pages 340, 341, and 353.
Answer: C

53. References: NFPA 1001, 5.3.6, 5.3.6(A)(B), 5.3.11, 5.3.11(A)(B), 5.3.12 and 5.3.12(A)(B)
Delmar, *Firefighter's Handbook*, 2nd Edition, 1st Printing, pages 396 and 398.
IFSTA, *Essentials of Fire Fighting*, 4th Edition, 1st Printing, page 293.
Jones and Bartlett, *Fundamentals of Fire Fighter Skills*, 1st Edition, 1st Printing, page 345.
Answer: B

54. References: NFPA 1001, 5.3.6, 5.3.6(A)(B), 5.3.11, 5.3.11(A)(B), 5.3.9 and 5.3.9(A)(B)
Delmar, *Firefighter's Handbook*, 2nd Edition, 1st Printing, pages 399-400.
IFSTA, *Essentials of Fire Fighting*, 4th Edition, 1st Printing, page 298.
Jones and Bartlett, *Fundamentals of Fire Fighter Skills*, 1st Edition, 1st Printing, page 352.
Answer: B

55. References: NFPA 1001, 5.3.6, 5.3.6(A)(B), 5.3.11, 5.3.11(A)(B), 5.3.11(A)(B), 5.3.12 and 5.3.12(A)(B)
Delmar, *Firefighter's Handbook*, 2nd Edition, 1st Printing, page 386.
IFSTA, *Essentials of Fire Fighting*, 4th Edition, 1st Printing, page 296.
Jones and Bartlett, *Fundamentals of Fire Fighter Skills*, 1st Edition, 1st Printing, pages 342-343.
Answer: B

56. References: NFPA 1001, 5.3.6, 5.3.6(A)(B), 5.3.9 and 5.3.9(A)(B)
Delmar, *Firefighter's Handbook*, 2nd Edition, 1st Printing, page 396.
IFSTA, *Essentials of Fire Fighting*, 4th Edition, 1st Printing, pages 292 and 296.
Jones and Bartlett, *Fundamentals of Fire Fighter Skills*, 1st Edition, 1st Printing, pages 342 and 343.
Answer: B

57. References: NFPA 1001, 5.3.6, 5.3.6(A)(B), 5.3.11, 5.3.11(A)(B), 5.3.12 and 5.3.12(A)(B)
Delmar, *Firefighter's Handbook*, 2nd Edition, 1st Printing, pages 384-385.
IFSTA, *Essentials of Fire Fighting*, 4th Edition, 1st Printing, page 291.
Jones and Bartlett, *Fundamentals of Fire Fighter Skills*, 1st Edition, 1st Printing, pages 342 and 352.
Answer: D

58. Reference: NFPA 5.3.7 and 5.3.7(A)(B)
Delmar, *Firefighter's Handbook*, 2nd Edition, 1st Printing, page 600.
IFSTA, *Essentials of Fire Fighting*, 4th Edition, 1st Printing, page 549.
Jones and Bartlett, *Fundamentals of Fire Fighter Skills*, 1st Edition, 1st Printing, page 637.
Answer: D

59. Reference: NFPA 5.3.7 and 5.3.7(A)(B)
Delmar, *Firefighter's Handbook*, 2nd Edition, 1st Printing, page 600.
IFSTA, *Essentials of Fire Fighting*, 4th Edition, 1st Printing, page 548.
Jones and Bartlett, *Fundamentals of Fire Fighter Skills*, 1st Edition, 1st Printing, page 638.
Answer: C

60. Reference: NFPA 1001, 5.3.7 and 5.3.7(A)(B)
Delmar, *Firefighter's Handbook*, 2nd Edition, 1st Printing, page 623.
IFSTA, *Essentials of Fire Fighting*, 4th Edition, 1st Printing, page 548.
Jones and Bartlett, *Fundamentals of Fire Fighter Skills*, 1st Edition, 1st Printing, page 637.
Answer: C

61. Reference: NFPA 1001, 5.3.8, 5.3.8(A), 5.3.10, and 5.3.10(A)
Delmar, *Firefighter's Handbook*, 2nd Edition, 1st Printing, page 227.
IFSTA, *Essentials of Fire Fighting*, 4th Edition, 1st Printing, pages 410-413.
Jones and Bartlett, *Fundamentals of Fire Fighter Skills*, 1st Edition, 1st Printing, page 630.
Answer: C

62. Reference: NFPA 1001, 5.3.8, 5.3.8(A)(B), 5.3.10 and 5.3.10(A)(B)
Delmar, *Firefighter's Handbook*, 2nd Edition, 1st Printing, pages 238-239, 241, 243, 244, 246, 247-248.
IFSTA, *Essentials of Fire Fighting*, 4th Edition, 1st Printing, page 478.
Jones and Bartlett, *Fundamentals of Fire Fighter Skills*, 1st Edition, 1st Printing, pages 503-512.
Answer: B

63. Reference: NFPA 1001, 5.3.8, 5.3.8(A), 5.3.10, and 5.3.10(A)
Delmar, *Firefighter's Handbook*, 2nd Edition, 1st Printing, page 222.
IFSTA, *Essentials of Fire Fighting*, 4th Edition, 1st Printing, page 398.
Jones and Bartlett, *Fundamentals of Fire Fighter Skills*, 1st Edition, 1st Printing, page 467.
Answer: C

64. Reference: NFPA 1001, 5.3.8 and 5.3.8(A)
Delmar, *Firefighter's Handbook*, 2nd Edition, 1st Printing, page 620.
IFSTA, *Essentials of Fire Fighting*, 4th Edition, 1st Printing, page 38.
Jones and Bartlett, *Fundamentals of Fire Fighter Skills*, 1st Edition, 1st Edition, page 634.
Answer: C

65. Reference: NFPA 1001, 5.3.8 and 5.3.8(A)
Delmar, *Firefighter's Handbook*, 2nd Edition, 1st Printing, page 170.
IFSTA, *Essentials of Fire Fighting*, 4th Edition, 1st Printing, page 177.
Jones and Bartlett, *Fundamentals of Fire Fighter Skills*, 1st Edition, 1st Edition, pages 378-381.
Answer: C

66. Reference: NFPA 1001, 5.3.9 and 5.3.9(A)(B)
Delmar, *Firefighter's Handbook*, 2nd Edition, 1st Printing, page 465.
IFSTA, *Essentials of Fire Fighting*, 4th Edition, 1st Printing, page 183.
Jones and Bartlett, *Fundamentals of Fire Fighter Skills*, 1st Edition, 1st Printing, page 414.
Answer: B

67. Reference: NFPA 1001, 5.3.9 and 5.3.9(A)(B)
Delmar, *Firefighter's Handbook*, 2nd Edition, 1st Printing, page 465.
IFSTA, *Essentials of Fire Fighting*, 4th Edition, 1st Printing, pages 176-177.
Jones and Bartlett, *Fundamentals of Fire Fighter Skills*, 1st Edition, 1st Printing, pages 378-379.
Answer: B

68. Reference: NFPA 1001, 5.3.9 and 5.3.9(A)(B)
Delmar, *Firefighter's Handbook*, 2nd Edition, 1st Printing, pages 469, 471, and 472.
IFSTA, *Essentials of Fire Fighting*, 4th Edition, 1st Printing, pages 228-229.
Jones and Bartlett, *Fundamentals of Fire Fighter Skills*, 1st Edition, 1st Printing, page 372.
Answer: C

69. Reference: NFPA 1001, 5.3.9 and 5.3.9(A)
Delmar, *Firefighter's Handbook*, 2nd Edition, 1st Printing, page 461.
IFSTA, *Essentials of Fire Fighting*, 4th Edition, 1st Printing, page 180.
Jones and Bartlett, *Fundamentals of Fire Fighter Skills*, 1st Edition, 1st Printing, page 375.
Answer: B

70. Reference: NFPA 1001, 5.3.10, 5.3.10(A)(B), 5.3.8 and 5.3.8(A)(B)
Delmar, *Firefighter's Handbook*, 2nd Edition, 1st Printing, pages 283-284.
IFSTA, *Essentials of Fire Fighting*, 4th Edition, 1st Printing, page 494.
Jones and Bartlett, *Fundamentals of Fire Fighter Skills*, 1st Edition, 1st Printing, page 626.
Answer: B

71. Reference: NFPA 1001, 5.3.10, 5.3.10(A)(B), 5.3.8 and 5.3.8(A)(B)
Delmar, *Firefighter's Handbook*, 2nd Edition, 1st Printing, pages 282-283.
IFSTA, *Essentials of Fire Fighting*, 4th Edition, 1st Printing, page 492.
Jones and Bartlett, *Fundamentals of Fire Fighter Skills*, 1st Edition, 1st Printing, page 626.
Answer: A

72. Reference: NFPA 1001, 5.3.10, 5.3.10(A)(B), 5.3.8 and 5.3.8(A)(B)
Delmar, *Firefighter's Handbook*, 2nd Edition, 1st Printing, pages 271 and 289.
IFSTA, *Essentials of Fire Fighting*, 4th Edition, 1st Printing, page 492.
Jones and Bartlett, *Fundamentals of Fire Fighter Skills*, 1st Edition, 1st Printing, page 520.
Answer: C

73. Reference: NFPA 1001, 5.3.10 and 5.3.10(A)(B)
Delmar, *Firefighter's Handbook*, 2nd Edition, 1st Printing, page 618.
IFSTA, *Essentials of Fire Fighting*, 4th Edition, 1st Printing, pages 545-546.
Jones and Bartlett, *Fundamentals of Fire Fighter Skills*, 1st Edition, 1st Printing, page 635.
Answer: B

74. Reference: NFPA 1001, 5.3.10, 5.3.10(A), 5.3.8 and 5.3.8(A)(B)
Delmar, *Firefighter's Handbook*, 2nd Edition, 1st Printing, pages 282-283.
IFSTA, *Essentials of Fire Fighting*, 4th Edition, 1st Printing, pages 492-496.
Jones and Bartlett, *Fundamentals of Fire Fighter Skills*, 1st Edition, 1st Printing, page 521.
Answer: A

75. Reference: NFPA 1001, 5.3.10, 5.3.10(A)(B), 5.3.8, and 5.3.8(A)(B)
Delmar, *Firefighter's Handbook*, 2nd Edition, 1st Printing, page 603.
IFSTA, *Essentials of Fire Fighting*, 4th Edition, 1st Printing, page 525.
Jones and Bartlett, *Fundamentals of Fire Fighter Skills*, 1st Edition, 1st Printing, page 635.
Answer: D

76. Reference: NFPA 1001, 5.3.10, 5.3.10(A)(B), 5.3.8 and 5.3.8(A)(B)
Delmar, *Firefighter's Handbook*, 2nd Edition, 1st Printing, page 604.
IFSTA, *Essentials of Fire Fighting*, 4th Edition, 1st Printing, page 527.
Jones and Bartlett, *Fundamentals of Fire Fighter Skills*, 1st Edition, 1st Printing, page 628.
Answer: C

77. Reference: NFPA 1001, 5.3.10, 5.3.10(A)(B), 5.3.8 and 5.3.8(A)(B)
Delmar, *Firefighter's Handbook*, 1st Edition, 2nd Printing, pages 283-284.
IFSTA, *Essentials of Fire Fighting*, 4th Edition, 1st Printing, page 496.
Jones and Bartlett, *Fundamentals of Fire Fighter Skills*, 1st Edition, 1st Printing, page 523.
Answer: D

78. Reference: NFPA 1001, 5.3.10, 5.3.10(A), 5.3.8 and 5.3.8(A)(B)
Delmar, *Firefighter's Handbook*, 2nd Edition, 1st Printing, page 283.
IFSTA, *Essentials of Fire Fighting*, 4th Edition, 1st Printing, page 493.
Jones and Bartlett, *Fundamentals of Fire Fighter Skills*, 1st Edition, 1st Printing, page 520.
Answer: A

79. Reference: NFPA 1001, 5.3.10, 5.3.10(A), 5.3.8 and 5.3.8(A)(B)
Delmar, *Firefighter's Handbook*, 2nd Edition, 1st Printing, page 282.
IFSTA, *Essentials of Fire Fighting*, 4th Edition, 1st Printing, page 487.
Jones and Bartlett, *Fundamentals of Fire Fighter Skills*, 1st Edition, 1st Printing, page 520.
Answer: D

80. Reference: NFPA 1001, 5.3.10, 5.3.10(A)(B), 5.3.8 and 5.3.8(A)(B)
Delmar, *Firefighter's Handbook*, 2nd Edition, 1st Printing, page 216.
IFSTA, *Essentials of Fire Fighting*, 4th Edition, 1st Printing, page 491.
Jones and Bartlett, *Fundamentals of Fire Fighter Skills*, 1st Edition, 1st Printing, pages 467 and 520.
Answer: B

81. Reference: NFPA 1001, 5.3.10, 5.3.10(A)(B), 5.3.8 and 5.3.8(A)(B)
Delmar, *Firefighter's Handbook*, 2nd Edition, 1st Printing, page 227.
IFSTA, *Essentials of Fire Fighting*, 4th Edition, 1st Printing, page 412.
Jones and Bartlett, *Fundamentals of Fire Fighter Skills*, 1st Edition, 1st Printing, pages 469-470.
Answer: D

82. Reference: NFPA 1001, 5.3.10, 5.3.10(A)(B), 5.3.8 and 5.3.8(A)(B)
Delmar, *Firefighter's Handbook*, 2nd Edition, 1st Printing, page 227.
IFSTA, *Essentials of Fire Fighting*, 4th Edition, 1st Printing, pages 410-411.
Jones and Bartlett, *Fundamentals of Fire Fighter Skills*, 1st Edition, 1st Printing, page 482.
Answer: C

83. Reference: NFPA 1001, 5.3.10 and 5.3.10(A)
Delmar, *Firefighter's Handbook*, 2nd Edition, 2nd Printing, page 236.
IFSTA, *Essentials of Fire Fighting*, 4th Edition, 1st Printing, pages 429 and 478.
Jones and Bartlett, *Fundamentals of Fire Fighter Skills*, 1st Edition, 1st Printing, page 511.
Answer: B

84. Reference: NFPA 1001, 5.3.10, 5.3.10(A)(B), 5.3.8 and 5.3.8(A)(B)
Delmar, *Firefighter's Handbook*, 2nd Edition, 2nd Printing, page 228.
IFSTA, *Essentials of Fire Fighting*, 4th Edition, 1st Printing, page 406.
Jones and Bartlett, *Fundamentals of Fire Fighter Skills*, 1st Edition, 1st Printing, page 479 and 532.
Answer: B

85. Reference: NFPA 1001, 5.3.10 and 5.3.10(A)(B)
Delmar, *Firefighter's Handbook*, 2nd Edition, 1st Printing, pages 255-257 and 616-618.
IFSTA, *Essentials of Fire Fighting*, 4th Edition, 1st Printing, page 431.
Jones and Bartlett, *Fundamentals of Fire Fighter Skills*, 1st Edition, 1st Printing, page 518.
Answer: D

86. Reference: NFPA 1001, 5.3.10, 5.3.10(A), 5.3.8 and 5.3.8(A)(B)
Delmar, *Firefighter's Handbook*, 2nd Edition, 1st Printing, page 225.
IFSTA, *Essentials of Fire Fighting*, 4th Edition, 1st Printing, page 402.
Jones and Bartlett, *Fundamentals of Fire Fighter Skills*, 1st Edition, 1st Printing, pages 469-470.
Answer: C

87. Reference: NFPA 1001, 5.3.11, 5.3.11(A), 5.3.12, 5.3.12(A), 5.3.13 and 5.3.13(A)
Delmar, *Firefighter's Handbook*, 2nd Edition, 1st Printing, page 76.
IFSTA, *Essentials of Fire Fighting*, 4th Edition, 1st Printing, page 40.
Jones and Bartlett, *Fundamentals of Fire Fighter Skills*, 1st Edition, 1st Printing, pages 124-125.
Answer: B

88. Reference: NFPA 1001, 5.3.11, 5.3.11(A), 5.3.12, 5.3.12(A), 5.3.13 and 5.3.13(A)
Delmar, *Firefighter's Handbook*, 2nd Edition, 1st Printing, page 86.
IFSTA, *Essentials of Fire Fighting*, 4th Edition, 1st Printing, page 42.
Jones and Bartlett, *Fundamentals of Fire Fighter Skills*, 1st Edition, 1st Printing, page 127.
Answer: B

89. Reference: NFPA 1001, 5.3.11, 5.3.11(A), 5.3.12, 5.3.12(A), 5.3.13 and 5.3.13(A)
Delmar, *Firefighter's Handbook*, 2nd Edition, 1st Printing, page 94.
IFSTA, *Essentials of Fire Fighting*, 4th Edition, 1st Printing, page 37.
Jones and Bartlett, *Fundamentals of Fire Fighter Skills*, 1st Edition, 1st Printing, page 130.
Answer: A

90. Reference: NFPA 1001, 5.3.11, 5.3.11(A), 5.3.12, 5.3.12(A), 5.3.13 and 5.3.13(A)
Delmar, *Firefighter's Handbook*, 2nd Edition, 1st Printing, pages 83 and 104.
IFSTA, *Essentials of Fire Fighting*, 4th Edition, 1st Printing, page 42.
Jones and Bartlett, *Fundamentals of Fire Fighter Skills*, 1st Edition, 1st Printing, pages 126, 128, 143, and 152.
Answer: B

91. Reference: NFPA 1001, 5.3.11, 5.3.11(A), 5.3.12, 5.3.12(A), 5.3.13 and 5.3.13(A)
Delmar, *Firefighter's Handbook*, 2nd Edition, 1st Printing, pages 92-93.
IFSTA, *Essentials of Fire Fighting*, 4th Edition, 1st Printing, page 49.
Jones and Bartlett, *Fundamentals of Fire Fighter Skills*, 1st Edition, 1st Printing, page 136.
Answer: A

92. Reference: NFPA 1001, 5.3.11, 5.3.11(A), 5.3.12, 5.3.12(A), 5.3.13 and 5.3.13(A)
Delmar, *Firefighter's Handbook*, 2nd Edition, 1st Printing, page 87.
IFSTA, *Essentials of Fire Fighting*, 4th Edition, 1st Printing, page 39.
Jones and Bartlett, *Fundamentals of Fire Fighter Skills*, 1st Edition, 1st Printing, page 133.
Answer: B

93. Reference: NFPA 1001, 5.3.11, 5.3.11(A), 5.3.12, 5.3.12(A), 5.3.13 and 5.3.13(A)
Delmar, *Firefighter's Handbook*, 2nd Edition, 1st Printing, pages 77-78.
IFSTA, *Essentials of Fire Fighting*, 4th Edition, 1st Printing, page 36.
Jones and Bartlett, *Fundamentals of Fire Fighter Skills*, 1st Edition, 1st Printing, page 129.
Answer: A

94. Reference: NFPA 1001, 5.3.11, 5.3.11(A), 5.3.12, 5.3.12(A), 5.3.13 and 5.3.13(A)
Delmar, *Firefighter's Handbook*, 2nd Edition, 1st Printing, pages 83 and 104.
IFSTA, *Essentials of Fire Fighting*, 4th Edition, 1st Printing, page 42.
Jones and Bartlett, *Fundamentals of Fire Fighter Skills*, 1st Edition, 1st Printing, pages 126, 128, and 152.
Answer: D

95. Reference: NFPA 1001, 5.3.11, 5.3.11(A), 5.3.10, 5.3.10(A), 5.3.12, 5.3.12(A), 5.3.13 and 5.3.13(A)
Delmar, *Firefighter's Handbook*, 2nd Edition, 1st Printing, page 85.
IFSTA, *Essentials of Fire Fighting*, 4th Edition, 1st Printing, page 40.
Jones and Bartlett, *Fundamentals of Fire Fighter Skills*, 1st Edition, 1st Printing, page 128.
Answer: C

96. Reference: NFPA 1001, 5.3.11, 5.3.11(A), 5.3.10, 5.3.10(A), 5.3.12, 5.3.12(A), 5.3.13 and 5.3.13(A)
Delmar, *Firefighter's Handbook*, 2nd Edition, 1st Printing, page 77.
IFSTA, *Essentials of Fire Fighting*, 4th Edition, 1st Printing, page 40.
Jones and Bartlett, *Fundamentals of Fire Fighter Skills*, 1st Edition, 1st Printing, page 124.
Answer: B

97. Reference: NFPA 1001, 5.3.11, 5.3.11(A), 5.3.10, 5.3.10(A), 5.3.12, and 5.3.12(A)
Delmar, *Firefighter's Handbook*, 2nd Edition, 1st Printing, page 556.
IFSTA, *Essentials of Fire Fighting*, 4th Edition, 1st Printing, pages 54-55.
Jones and Bartlett, *Fundamentals of Fire Fighter Skills*, 1st Edition, 1st Printing, pages 139, 305, 410, 418, and 439.
Answer: C

98. Reference: NFPA 1001, 5.3.11, 5.3.11(A), 5.3.12, and 5.3.12(A)
Delmar, *Firefighter's Handbook*, 2nd Edition, 1st Printing, pages 80 and 104.
IFSTA, *Essentials of Fire Fighting*, 4th Edition, 1st Printing, page 42.
Jones and Bartlett, *Fundamentals of Fire Fighter Skills*, 1st Edition, 1st Printing, page 805.
Answer: C

99. Reference: NFPA 1001, 5.3.11, 5.3.11(A), 5.3.10, 5.3.10(A), 5.3.12, and 5.3.12(A)
Delmar, *Firefighter's Handbook*, 2nd Edition, 1st Printing, page 83.
IFSTA, *Essentials of Fire Fighting*, 4th Edition, 1st Printing, page 42.
Jones and Bartlett, *Fundamentals of Fire Fighter Skills*, 1st Edition, 1st Printing, page 126.
Answer: D

100. Reference: NFPA 1001, 5.3.11, 5.3.11(A), 5.3.12, and 5.3.12(A)
Delmar, *Firefighter's Handbook*, 2nd Edition, 1st Printing, pages 87-88.
IFSTA, *Essentials of Fire Fighting*, 4th Edition, 1st Printing, page 39.
Jones and Bartlett, *Fundamentals of Fire Fighter Skills*, 1st Edition, 1st Printing, page 133.
Answer: D

101. Reference: NFPA 1001, 5.3.12 and 5.3.12(A)(B)
Delmar, *Firefighter's Handbook*, 2nd Edition, 1st Printing, page 410.
IFSTA, *Essentials of Fire Fighting*, 4th Edition, 1st Printing, pages 372 and 374.
Jones and Bartlett, *Fundamentals of Fire Fighter Skills*, 1st Edition, 1st Printing, page 424.
Answer: B

102. Reference: NFPA 1001, 5.3.12 and 5.3.12(A)(B)
Delmar, *Firefighter's Handbook*, 2nd Edition, 1st Printing, page 562.
IFSTA, *Essentials of Fire Fighting*, 4th Edition, 1st Printing, page 348.
Jones and Bartlett, *Fundamentals of Fire Fighter Skills*, 1st Edition, 1st Printing, pages 140, 410, and 423.
Answer: C

103. Reference: NFPA 1001, 5.3.12 and 5.3.12(A)(B)
Delmar, *Firefighter's Handbook*, 2nd Edition, 1st Printing, pages 575 and 577.
IFSTA, *Essentials of Fire Fighting*, 4th Edition, 1st Printing, pages 358 and 374.
Jones and Bartlett, *Fundamentals of Fire Fighter Skills*, 1st Edition, 1st Printing, pages 430-431.
Answer: C

104. Reference: NFPA 1001, 5.3.12 and 5.3.12(A)(B)
Delmar, *Firefighter's Handbook*, 2nd Edition, 1st Printing, pages 579-581.
IFSTA, *Essentials of Fire Fighting*, 4th Edition, 1st Printing, page 361.
Jones and Bartlett, *Fundamentals of Fire Fighter Skills*, 1st Edition, 1st Printing, pages 434-436.
Answer: A

105. Reference: NFPA 1001, 5.3.12 and 5.3.12(A)(B)
Delmar, *Firefighter's Handbook*, 2nd Edition, 1st Printing, page 572.
IFSTA, *Essentials of Fire Fighting*, 4th Edition, 1st Printing, page 356.
Jones and Bartlett, *Fundamentals of Fire Fighter Skills*, 1st Edition, 1st Printing, pages 423-424.
Answer: B

106. Reference: NFPA 1001, 5.3.12, 5.3.12(A)(B), 5.3.11 and 5.3.11(A)(B)
Delmar, *Firefighter's Handbook*, 2nd Edition, 1st Printing, page 551.
IFSTA, *Essentials of Fire Fighting*, 4th Edition, 1st Printing, page 347.
Jones and Bartlett, *Fundamentals of Fire Fighter Skills*, 1st Edition, 1st Printing, page 409.
Answer: C

107. Reference: NFPA 1001, 5.3.12, 5.3.12(A)(B), 5.3.11 and 5.3.11(A)(B)
Delmar, *Firefighter's Handbook*, 2nd Edition, 1st Printing, page 563.
IFSTA, *Essentials of Fire Fighting*, 4th Edition, 1st Printing, page 366.
Jones and Bartlett, *Fundamentals of Fire Fighter Skills*, 1st Edition, 1st Printing, pages 414-419.
Answer: A

108. Reference: NFPA 1001, 5.3.12, 5.3.12(A), 5.3.10, 5.3.10(A), 5.3.11 and 5.3.11(A)
Delmar, *Firefighter's Handbook*, 2nd Edition, 1st Printing, page 349.
IFSTA, *Essentials of Fire Fighting*, 4th Edition, 1st Printing, page 70.
Jones and Bartlett, *Fundamentals of Fire Fighter Skills*, 1st Edition, 1st Printing, page 150.
Answer: D

109. Reference: NFPA 1001, 5.3.12, 5.3.12(A), 5.3.11 and 5.3.11(A)
Delmar, *Firefighter's Handbook*, 2nd Edition, 1st Printing, page 349.
IFSTA, *Essentials of Fire Fighting*, 4th Edition, 1st Printing, page 70.
Jones and Bartlett, *Fundamentals of Fire Fighter Skills*, 1st Edition, 1st Printing, page 150.
Answer: B

110. Reference: NFPA 1001, 5.3.12, 5.3.12(A), 5.3.10, 5.3.10(A), 5.3.11 and 5.3.11(A)
Delmar, *Firefighter's Handbook*, 2nd Edition, 1st Printing, pages 361-362.
IFSTA, *Essentials of Fire Fighting*, 4th Edition, 1st Printing, page 75.
Jones and Bartlett, *Fundamentals of Fire Fighter Skills*, 1st Edition, 1st Printing, page 427.
Answer: A

111. Reference: NFPA 1001, 5.3.13 and 5.3.13(A)(B)
Delmar, *Firefighter's Handbook*, 2nd Edition, 1st Printing, page 649.
IFSTA, *Essentials of Fire Fighting*, 4th Edition, 1st Printing, page 596.
Jones and Bartlett, *Fundamentals of Fire Fighter Skills*, 1st Edition, 1st Printing, page 574.
Answer: A

112. Reference: NFPA 1001, 5.3.13 and 5.3.13(A)(B)
Delmar, *Firefighter's Handbook*, 2nd Edition, 1st Printing, pages 649-651.
IFSTA, *Essentials of Fire Fighting*, 4th Edition, 1st Printing, page 598.
Jones and Bartlett, *Fundamentals of Fire Fighter Skills*, 1st Edition, 1st Printing, page 577.
Answer: C

113. Reference: NFPA 1001, 5.3.13 and 5.3.13(A)(B)
Delmar, *Firefighter's Handbook*, 2nd Edition, 1st Printing, page 649.
IFSTA, *Essentials of Fire Fighting*, 4th Edition, 1st Printing, page 597.
Jones and Bartlett, *Fundamentals of Fire Fighter Skills*, 1st Edition, 1st Printing, page 576.
Answer: C

114. Reference: NFPA 1001, 5.3.13 and 5.3.13(A)
Delmar, *Firefighter's Handbook*, 2nd Edition, 1st Printing, page 635.
IFSTA, *Essentials of Fire Fighting*, 4th Edition, 1st Printing, page 593.
Jones and Bartlett, *Fundamentals of Fire Fighter Skills*, 1st Edition, 1st Printing, page 571.
Answer: A

115. Reference: NFPA 1001, 5.3.14 and 5.3.14(A)(B)
Delmar, *Firefighter's Handbook*, 2nd Edition, 1st Printing, page 646.
IFSTA, *Essentials of Fire Fighting*, 4th Edition, 1st Printing, page 595.
Jones and Bartlett, *Fundamentals of Fire Fighter Skills*, 1st Edition, 1st Printing, pages 569-572.
Answer: A

116. Reference: NFPA 1001, 5.3.14 and 5.3.14(A)(B)
Delmar, *Firefighter's Handbook*, 2nd Edition, 1st Printing, page 646.
IFSTA, *Essentials of Fire Fighting*, 4th Edition, 1st Printing, pages 615-616.
Jones and Bartlett, *Fundamentals of Fire Fighter Skills*, 1st Edition, 1st Printing, page 571.
Answer: B

117. Reference: NFPA 1001, 5.3.14 and 5.3.14(A)(B)
Delmar, *Firefighter's Handbook*, 2nd Edition, 1st Printing, page 633.
IFSTA, *Essentials of Fire Fighting*, 4th Edition, 1st Printing, page 587.
Jones and Bartlett, *Fundamentals of Fire Fighter Skills*, 1st Edition, 1st Printing, page 554.
Answer: C

118. Reference: NFPA 1001, 5.3.14 and 5.3.14(A)(B)
Delmar, *Firefighter's Handbook*, 2nd Edition, 1st Printing, page 639.
IFSTA, *Essentials of Fire Fighting*, 4th Edition, 1st Printing, page 587.
Jones and Bartlett, *Fundamentals of Fire Fighter Skills*, 1st Edition, 1st Printing, page 555.
Answer: B

119. Reference: NFPA 1001, 5.3.14 and 5.3.14(A)(B)
Delmar, *Firefighter's Handbook*, 2nd Edition, 1st Printing, page 643.
Jones and Bartlett, *Fundamentals of Fire Fighter Skills*, 1st Edition, 1st Printing, pages 560 and 563.
Answer: D

120. Reference: NFPA 1001, 5.3.14 and 5.3.14(A)
Delmar, *Firefighter's Handbook*, 2nd Edition, 1st Printing, page 331.
IFSTA, *Essentials of Fire Fighting*, 4th Edition, 1st Printing, page 578.
Jones and Bartlett, *Fundamentals of Fire Fighter Skills*, 1st Edition, 1st Printing, page 943.
Answer: C

121. Reference: NFPA 1001, 5.3.15 and 5.3.15(A)(B)
Delmar, *Firefighter's Handbook*, 2nd Edition, 1st Printing, pages 210-211.
IFSTA, *Essentials of Fire Fighting*, 4th Edition, 1st Printing, page 392.
Jones and Bartlett, *Fundamentals of Fire Fighter Skills*, 1st Edition, 1st Printing, pages 458-459.
Answer: A

122. Reference: NFPA 1001, 5.3.15 and 5.3.15(A)
Delmar, *Firefighter's Handbook*, 2nd Edition, 1st Printing, pages 210-211.
IFSTA, *Essentials of Fire Fighting*, 4th Edition, 1st Printing, page 379.
Jones and Bartlett, *Fundamentals of Fire Fighter Skills*, 1st Edition, 1st Printing, page 446.
Answer: B

123. Reference: NFPA 1001, 5.3.15 and 5.3.15(A)
Delmar, *Firefighter's Handbook*, 2nd Edition, 1st Printing, pages 291-292.
IFSTA, *Essentials of Fire Fighting*, 4th Edition, 1st Printing, pages 385-386.
Jones and Bartlett, *Fundamentals of Fire Fighter Skills*, 1st Edition, 1st Printing, pages 454-457.
Answer: D

124. Reference: NFPA 1001, 5.3.15 and 5.3.15(A)(B)
Delmar, *Firefighter's Handbook*, 2nd Edition, 1st Printing, page 205.
IFSTA, *Essentials of Fire Fighting*, 4th Edition, 1st Printing, page 391.
Jones and Bartlett, *Fundamentals of Fire Fighter Skills*, 1st Edition, 1st Printing, page 458.
Answer: C

125. Reference: NFPA 1001, 5.3.15 and 5.3.15(A)
Delmar, *Firefighter's Handbook*, 2nd Edition, 1st Printing, page 205.
IFSTA, *Essentials of Fire Fighting*, 4th Edition, 1st Printing, page 381.
Jones and Bartlett, *Fundamentals of Fire Fighter Skills*, 1st Edition, 1st Printing, page 448.
Answer: B

126. Reference: NFPA 1001, 5.3.15 and 5.3.15(A)
Delmar, *Firefighter's Handbook*, 2nd Edition, 1st Printing, pages 206-207.
IFSTA, *Essentials of Fire Fighting*, 4th Edition, 1st Printing, page 386.
Jones and Bartlett, *Fundamentals of Fire Fighter Skills*, 1st Edition, 1st Printing, pages 449-450.
Answer: A

127. Reference: NFPA 1001, 5.3.16 and 5.3.16(A)
Delmar, *Firefighter's Handbook*, 2nd Edition, 1st Printing, page 186.
IFSTA, *Essentials of Fire Fighting*, 4th Edition, 1st Printing, pages 126, 132, and 135.
Jones and Bartlett, *Fundamentals of Fire Fighter Skills*, 1st Edition, 1st Printing, pages 188 and 193.
Answer: A

128. Reference: NFPA 1001, 5.3.16 and 5.3.16(A)(B)
Delmar, *Firefighter's Handbook*, 2nd Edition, 1st Printing, pages 185-186.
IFSTA, *Essentials of Fire Fighting*, 4th Edition, 1st Printing, page 58.
Jones and Bartlett, *Fundamentals of Fire Fighter Skills*, 1st Edition, 1st Printing, page 178.
Answer: B

APPENDIX A. EXAMINATION I-2 ANSWER KEY

129. Reference: NFPA 1001, 5.3.16 and 5.3.16(A)
Delmar, *Firefighter's Handbook*, 2nd Edition, 1st Printing, page 189.
IFSTA, *Essentials of Fire Fighting*, 4th Edition, 1st Printing, pages 133-135.
Jones and Bartlett, *Fundamentals of Fire Fighter Skills*, 1st Edition, 1st Printing, page 181.
Answer: B

130. Reference: NFPA 1001, 5.3.16 and 5.3.16(A)
Delmar, *Firefighter's Handbook*, 2nd Edition, 1st Printing, pages 188 and 192.
IFSTA, *Essentials of Fire Fighting*, 4th Edition, 1st Printing, page 130.
Jones and Bartlett, *Fundamentals of Fire Fighter Skills*, 1st Edition, 1st Printing, page 186.
Answer: B

131. Reference: NFPA 1001, 5.3.16 and 5.3.16(A)(B)
Delmar, *Firefighter's Handbook*, 2nd Edition, 1st Printing, pages 193-194.
IFSTA, *Essentials of Fire Fighting*, 4th Edition, 1st Printing, page 134.
Jones and Bartlett, *Fundamentals of Fire Fighter Skills*, 1st Edition, 1st Printing, page 180.
Answer: A

132. Reference: NFPA 1001, 5.3.16 and 5.3.16(A)
Delmar, *Firefighter's Handbook*, 2nd Edition, 1st Printing, page 189.
IFSTA, *Essentials of Fire Fighting*, 4th Edition, 1st Printing, page 135.
Jones and Bartlett, *Fundamentals of Fire Fighter Skills*, 1st Edition, 1st Printing, page 181.
Answer: A

133. Reference: NFPA 1001, 5.3.16 and 5.3.16(A)
Delmar, *Firefighter's Handbook*, 2nd Edition, 1st Printing, pages 186-187.
Jones and Bartlett, *Fundamentals of Fire Fighter Skills*, 1st Edition, 1st Printing, pages 179-180.
Answer: D

134. Reference: NFPA 1001, 5.3.18 and 5.3.18(A)(B)
Delmar, *Firefighter's Handbook*, 2nd Edition, 1st Printing, page 499.
IFSTA, *Essentials of Fire Fighting*, 4th Edition, 1st Printing, page 538.
Jones and Bartlett, *Fundamentals of Fire Fighter Skills*, 1st Edition, 1st Printing, page 642.
Answer: C

135. Reference: NFPA 1001, 5.3.19 and 5.3.19(A)
Delmar, *Firefighter's Handbook*, 2nd Edition, 1st Printing, pages 599 and 628.
IFSTA, *Essentials of Fire Fighting*, 4th Edition, 1st Printing, page 554.
Answer: A

136. Reference: NFPA 1001, 5.3.19 and 5.3.19(A)
Delmar, *Firefighter's Handbook*, 2nd Edition, 1st Printing, pages 595-599.
IFSTA, *Essentials of Fire Fighting*, 4th Edition, 1st Printing, page 554.
Jones and Bartlett, *Fundamentals of Fire Fighter Skills*, 1st Edition, 1st Printing, pages 609-611.
Answer: C

137. Reference: NFPA 1001, 5.5.1 and 5.5.1(A)(B)
Delmar, *Firefighter's Handbook*, 2nd Edition, 1st Printing, pages 678-679.
IFSTA, *Essentials of Fire Fighting*, 4th Edition, 1st Printing, page 667.
Jones and Bartlett, *Fundamentals of Fire Fighter Skills*, 1st Edition, 1st Printing, page 912.
Answer: D

138. Reference: NFPA 1001, 5.5.1 and 5.5.1(A)(B)
Delmar, *Firefighter's Handbook*, 2nd Edition, 1st Printing, page 661.
IFSTA, *Essentials of Fire Fighting*, 4th Edition, 1st Printing, page 661.
Jones and Bartlett, *Fundamentals of Fire Fighter Skills*, 1st Edition, 1st Printing, page 651.
Answer: C

139. Reference: NFPA 1001, 5.5.1 and 5.5.1(A)(B)
Delmar, *Firefighter's Handbook*, 2nd Edition, 1st Printing, page 666.
IFSTA, *Essentials of Fire Fighting*, 4th Edition, 1st Printing, pages 137 and 138.
Jones and Bartlett, *Fundamentals of Fire Fighter Skills*, 1st Edition, 1st Printing, page 207.
Answer: B

140. Reference: NFPA 1001, 5.5.1 and 5.5.1(A)(B)
Delmar, *Firefighter's Handbook*, 2nd Edition, 1st Printing, page 685.
IFSTA, *Essentials of Fire Fighting*, 4th Edition, 1st Printing, page 662.
Jones and Bartlett, *Fundamentals of Fire Fighter Skills*, 1st Edition, 1st Printing, page 648.
Answer: D

141. Reference: NFPA 1001, 5.5.1 and 5.5.1(A)
Delmar, *Firefighter's Handbook*, 2nd Edition, 1st Printing, page 685.
IFSTA, *Essentials of Fire Fighting*, 4th Edition, 1st Printing, page 664.
Jones and Bartlett, *Fundamentals of Fire Fighter Skills*, 1st Edition, 1st Printing, page 656.
Answer: B

142. Reference: NFPA 1001, 5.5.2 and 5.5.2(A)(B)
Delmar, *Firefighter's Handbook*, 2nd Edition, 1st Printing, page 680.
IFSTA, *Essentials of Fire Fighting*, 4th Edition, 1st Printing, page 671.
Jones and Bartlett, *Fundamentals of Fire Fighter Skills*, 1st Edition, 1st Printing, page 917.
Answer: D

143. Reference: NFPA 1001, 5.5.3 and 5.5.3(A)(B)
Delmar, *Firefighter's Handbook*, 2nd Edition, 1st Printing, page 177.
IFSTA, *Essentials of Fire Fighting*, 4th Edition, 1st Printing, page 113.
Jones and Bartlett, *Fundamentals of Fire Fighter Skills*, 1st Edition, 1st Printing, page 62.
Answer: B

APPENDIX A. EXAMINATION I-2 ANSWER KEY

144. Reference: NFPA 1001, 5.5.3 and 5.5.3(A)(B)
Delmar, *Firefighter's Handbook*, 2nd Edition, 1st Printing, page 155.
IFSTA, *Essentials of Fire Fighting*, 4th Edition, 1st Printing, page 106.
Jones and Bartlett, *Fundamentals of Fire Fighter Skills*, 1st Edition, 1st Printing, page 62.
Answer: B

145. Reference: NFPA 1001, 5.5.3 and 5.5.3(A)(B)
Delmar, *Firefighter's Handbook*, 2nd Edition, 1st Printing, page 172.
IFSTA, *Essentials of Fire Fighting*, 4th Edition, 1st Printing, page 106.
Jones and Bartlett, *Fundamentals of Fire Fighter Skills*, 1st Edition, 1st Printing, page 62.
Answer: C

146. Reference: NFPA 1001, 5.5.3 and 5.5.3(A)(B)
Delmar, *Firefighter's Handbook*, 2nd Edition, 1st Printing, page 154.
IFSTA, *Essentials of Fire Fighting*, 4th Edition, 1st Printing, page 106.
Jones and Bartlett, *Fundamentals of Fire Fighter Skills*, 1st Edition, 1st Printing, page 62.
Answer: B

147. Reference: NFPA 1001, 5.5.3 and 5.5.3(A)
Delmar, *Firefighter's Handbook*, 2nd Edition, 1st Printing, page 381.
IFSTA, *Essentials of Fire Fighting*, 4th Edition, 1st Printing, page 288.
Jones and Bartlett, *Fundamentals of Fire Fighter Skills*, 1st Edition, 1st Printing, page 339.
Answer: C

148. Reference: NFPA 1001, 5.5.4 and 5.5.4(A)(B)
Delmar, *Firefighter's Handbook*, 2nd Edition, 1st Printing, page 223.
IFSTA, *Essentials of Fire Fighting*, 4th Edition, 1st Printing, page 399.
Jones and Bartlett, *Fundamentals of Fire Fighter Skills*, 1st Edition, 1st Printing, pages 467 and 478.
Answer: B

149. Reference: NFPA 1001, 5.5.4 and 5.5.4(A)(B)
Delmar, *Firefighter's Handbook*, 2nd Edition, 1st Printing, page 223.
IFSTA, *Essentials of Fire Fighting*, 4th Edition, 1st Printing, page 399.
Jones and Bartlett, *Fundamentals of Fire Fighter Skills*, 1st Edition, 1st Printing, page 478.
Answer: C

150. Reference: NFPA 1001, 5.5.4 and 5.5.4(A)(B)
Delmar, *Firefighter's Handbook*, 2nd Edition, 1st Printing, page 224.
IFSTA, *Essentials of Fire Fighting*, 4th Edition, 1st Printing, page 400.
Jones and Bartlett, *Fundamentals of Fire Fighter Skills*, 1st Edition, 1st Printing, page 479.
Answer: C

Examination I-3 Answer Key

Directions
Follow these steps carefully for completing the feedback step of Systematic Approach to Exam Preparation (SAEP):

1. After entering your scores, look up the answers for the examination items you missed as well as those you guessed, even if you guessed correctly. If you are guessing, it means the answer is not perfectly clear. In this process we are committed to making you as knowledgeable as possible.
2. Enter the number of missed and guessed examination items in the blank on the personal progress plotter.
3. Highlight the answer in the reference materials. Once you have highlighted the answer, read the paragraph preceding and the paragraph following the one in which the correct answer is located. Now that you have highlighted the answer, enter the paragraph number and page number next to the guessed or missed examination item on your examination. Count any part of a paragraph at the beginning of the page as one paragraph until you reach the paragraph containing your highlighted answer. This step will help you locate and review your missed and guessed examination items later in the process. This step is __essential__ to learning the material in context and by association. These learning techniques (context/association) are the very backbone of the SAEP approach.
4. **Congratulations!** You have completed the Examination and feedback steps of SAEP when you have highlighted your guessed and missed examination items for this examination.

Proceed to Phase III and Phase IV. Study the materials carefully in these important phases. They will help you polish your examination taking skills. Approximately two to three days **prior** to taking your next examination, carefully read all the highlighted information in the reference materials using the same techniques applied during the feedback step. This will reinforce your learning and provide you with an added level of confidence going into the examination.

Someone once said to professional golfer, Tom Watson, after he won several tournament championships, "You are really lucky to have won those championships. You are really on a streak." Tom was reported to have said, "Yes, there is some luck involved, but what I have really noticed is that the more I practice the luckier I get." What Watson was saying is that good luck usually is the result of good preparation. This line of thinking certainly applies to learning the rules and hints of examination taking.

Rule 7

Good Luck = Good Preparation.

APPENDIX A. EXAMINATION I-3 ANSWER KEY

1. Reference: NFPA 1001, 5.1.1.1
Delmar, *Firefighter's Handbook*, 2nd Edition, 1st Printing, page 35.
IFSTA, *Essentials of Fire Fighting*, 4th Edition, 1st Printing, page 12.
Jones and Bartlett, *Fundamentals of Fire Fighter Skills*, 1st Edition, 1st Printing, page 13.
Answer: D

2. Reference: NFPA 1001, 5.1.1.1
Delmar, *Firefighter's Handbook*, 2nd Edition, 1st Printing, page 114.
Jones and Bartlett, *Fundamentals of Fire Fighter Skills*, 1st Edition, 1st Printing, pages 24-25.
Answer: A

3. Reference: NFPA 1001, 5.1.1.1
Delmar, *Firefighter's Handbook*, 1st Edition, 1st Printing, pages 112-114.
IFSTA, *Essentials of Fire Fighting*, 4th Edition, 1st Printing, page 26.
Jones and Bartlett, *Fundamentals of Fire Fighter Skills*, 1st Edition, 1st Printing, page 25.
Answer: D

4. Reference: NFPA 1001, 5.1.1.1
Delmar, *Firefighter's Handbook*, 2nd Edition, 1st Printing, pages 24-25.
IFSTA, *Essentials of Fire Fighting*, 4th Edition, 1st Printing, page 6.
Jones and Bartlett, *Fundamentals of Fire Fighter Skills*, 1st Edition, 1st Printing, pages 8-9.
Answer: D

5. Reference: NFPA 1001, 5.1.1.1
Delmar, *Firefighter's Handbook*, 2nd Edition, 1st Printing, page 37.
IFSTA, *Essentials of Fire Fighting*, 4th Edition, 1st Printing, page 14.
Jones and Bartlett, *Fundamentals of Fire Fighter Skills*, 1st Edition, 1st Printing, pages 24, 100-121.
Answer: D

6. Reference: NFPA 1001, 5.1.1.1, and 5.1.1.2
Delmar, *Firefighter's Handbook*, 2nd Edition, 1st Printing, page 427.
IFSTA, *Essentials of Fire Fighting*, 4th Edition, 1st Printing, pages 154, 155, and 165.
Jones and Bartlett, *Fundamentals of Fire Fighter Skills*, 1st Edition, 1st Printing, page 257.
Answer: A

7. Reference: NFPA 1001, 5.1.1.1, and 5.1.1.2
Delmar, *Firefighter's Handbook*, 2nd Edition, 1st Printing, page 425.
IFSTA, *Essentials of Fire Fighting*, 4th Edition, 1st Printing, page 154.
Jones and Bartlett, *Fundamentals of Fire Fighter Skills*, 1st Edition, 1st Printing, page 254.
Answer: C

8. Reference: NFPA 1001, 5.1.1.1, and 5.1.1.2
   Delmar, *Firefighter's Handbook*, 2nd Edition, 1st Printing, page 455.
   IFSTA, *Essentials of Fire Fighting*, 4th Edition, 1st Printing, page 157.
   Jones and Bartlett, *Fundamentals of Fire Fighter Skills*, 1st Edition, 1st Printing, page 268.
   Answer: C

9. Reference: NFPA 1001, 5.1.1.1, and 5.1.1.2
   Delmar, *Firefighter's Handbook*, 2nd Edition, 1st Printing, page 426.
   IFSTA, *Essentials of Fire Fighting*, 4th Edition, 1st Printing, page 155.
   Jones and Bartlett, *Fundamentals of Fire Fighter Skills*, 1st Edition, 1st Printing, page 255.
   Answer: B

10. Reference: NFPA 1001, 5.1.1.1 and 5.1.1.2
    Delmar, *Firefighter's Handbook*, 2nd Edition, 1st Printing, page 429.
    IFSTA, *Essentials of Fire Fighting*, 4th Edition, 1st Printing, pages 154 and 165.
    Jones and Bartlett, *Fundamentals of Fire Fighter Skills*, 1st Edition, 1st Printing, page 259.
    Answer: C

11. Reference: NFPA 1001, 5.1.1.1, and 5.1.1.2
    Delmar, *Firefighter's Handbook*, 2nd Edition, 1st Printing, page 429.
    IFSTA, *Essentials of Fire Fighting*, 4th Edition, 1st Printing, page 169.
    Jones and Bartlett, *Fundamentals of Fire Fighter Skills*, 1st Edition, 1st Printing, page 262.
    Answer: B

12. Reference: NFPA 1001, 5.1.1.1
    Delmar, *Firefighter's Handbook*, 2nd Edition, 1st Printing, page 426.
    IFSTA, *Essentials of Fire Fighting*, 4th Edition, 1st Printing, page 155.
    Jones and Bartlett, *Fundamentals of Fire Fighter Skills*, 1st Edition, 1st Printing, pages 254-255.
    Answer: C

13. Reference: NFPA 1001, 5.1.1.1
    Delmar, *Firefighter's Handbook*, 2nd Edition, 1st Printing, pages 442-443.
    IFSTA, *Essentials of Fire Fighting*, 4th Edition, 1st Printing, page 153.
    Jones and Bartlett, *Fundamentals of Fire Fighter Skills*, 1st Edition, 1st Printing, page 252.
    Answer: B

14. Reference: NFPA 1001, 5.1.1.1 and 5.1.1.2
    Delmar, *Firefighter's Handbook*, 2nd Edition, 1st Printing, pages 450-451.
    IFSTA, *Essentials of Fire Fighting*, 4th Edition, 1st Printing, page 157.
    Jones and Bartlett, *Fundamentals of Fire Fighter Skills*, 1st Edition, 1st Printing, pages 267-268.
    Answer: C

APPENDIX A. EXAMINATION I-3 ANSWER KEY

15. Reference: NFPA 1001, 5.1.1.1
Delmar, *Firefighter's Handbook*, 2nd Edition, 1st Printing, page 440.
IFSTA, *Essentials of Fire Fighting*, 4th Edition, 1st Printing, page 151.
Jones and Bartlett, *Fundamentals of Fire Fighter Skills*, 1st Edition, 1st Printing, page 252.
Answer: A

16. Reference: NFPA 1001, 5.2.1 and 5.2.1(A)(B)
Delmar, *Firefighter's Handbook*, 2nd Edition, 1st Printing, page 63.
IFSTA, *Essentials of Fire Fighting*, 4th Edition, 1st Printing, page 645.
Jones and Bartlett, *Fundamentals of Fire Fighter Skills*, 1st Edition, 1st Printing, page 83.
Answer: C

17. Reference: NFPA 1001, 5.2.1 and 5.2.1(A)
Delmar, *Firefighter's Handbook*, 2nd Edition, 1st Printing, pages 60-62.
IFSTA, *Essentials of Fire Fighting*, 4th Edition, 1st Printing, page 644.
Jones and Bartlett, *Fundamentals of Fire Fighter Skills*, 1st Edition, 1st Printing, page 81.
Answer: B

18. Reference: NFPA 1001, 5.2.1 and 5.2.1(A)(B)
Delmar, *Firefighter's Handbook*, 2nd Edition, 1st Printing, page 48.
IFSTA, *Essentials of Fire Fighting*, 4th Edition, 1st Printing, page 635.
Jones and Bartlett, *Fundamentals of Fire Fighter Skills*, 1st Edition, 1st Printing, page 73.
Answer: B

19. Reference: NFPA 1001, 5.2.3 and 5.2.3(A)(B)
Delmar, *Firefighter's Handbook*, 2nd Edition, 1st Printing, pages 65-66.
IFSTA, *Essentials of Fire Fighting*, 4th Edition, 1st Printing, page 647.
Jones and Bartlett, *Fundamentals of Fire Fighter Skills*, 1st Edition, 1st Printing, page 85.
Answer: C

20. Reference: NFPA 1001, 5.3.1, 5.3.1(A)(B), 5.3.5(A)(B), and 5.3.9(A)(B)
Delmar, *Firefighter's Handbook*, 2nd Edition, 1st Printing, page 156.
IFSTA, *Essentials of Fire Fighting*, 4th Edition, 1st Printing, page 97.
Jones and Bartlett, *Fundamentals of Fire Fighter Skills*, 1st Edition, 1st Printing, page 48.
Answer: B

21. Reference: NFPA 1001, 5.3.1, 5.3.1(A)(B), and 5.1.1.2
Delmar, *Firefighter's Handbook*, 2nd Edition, 1st Printing, pages 166-167.
IFSTA, *Essentials of Fire Fighting*, 4th Edition, 1st Printing, page 104.
Jones and Bartlett, *Fundamentals of Fire Fighter Skills*, 1st Edition, 1st Printing, pages 56 and 58.
Answer: A

22. Reference: NFPA 1001, 5.3.1, 5.3.1(A)(B), and 5.3.5(A)(B)
Delmar, *Firefighter's Handbook*, 2nd Edition, 1st Printing, page 145.
IFSTA, *Essentials of Fire Fighting*, 4th Edition, 1st Printing, page 88.
Jones and Bartlett, *Fundamentals of Fire Fighter Skills*, 1st Edition, 1st Printing, page 43.
Answer: D

23. Reference: NFPA 1001, 5.3.1, 5.3.1(A), 5.3.5(A)(B), 5.3.9, and 5.3.9(A)(B)
Delmar, *Firefighter's Handbook*, 2nd Edition, 1st Printing, pages 145 and 498.
IFSTA, *Essentials of Fire Fighting*, 4th Edition, 1st Printing, pages 91-92.
Jones and Bartlett, *Fundamentals of Fire Fighter Skills*, 1st Edition, 1st Printing, pages 42-44.
Answer: D

24. Reference: NFPA 1001, 5.3.1, 5.3.1(A)(B), 5.3.5(A)(B), 5.3.9(A)(B), and 5.3.10(A)
Delmar, *Firefighter's Handbook*, 2nd Edition, 1st Printing, pages 135-136, 144.
IFSTA, *Essentials of Fire Fighting*, 4th Edition, 1st Printing, pages 99-100.
Jones and Bartlett, *Fundamentals of Fire Fighter Skills*, 1st Edition, 1st Printing, pages 35-36, 67.
Answer: B

25. Reference: NFPA 1001, 5.3.1, 5.3.1(A), and 5.3.5(A)(B)
Delmar, *Firefighter's Handbook*, 2nd Edition, 1st Printing, page 156.
IFSTA, *Essentials of Fire Fighting*, 4th Edition, 1st Printing, page 97.
Jones and Bartlett, *Fundamentals of Fire Fighter Skills*, 1st Edition, 1st Printing, page 48.
Answer: C

26. Reference: NFPA 1001, 5.3.1, 5.3.1(A)(B), 5.3.5, 5.3.5(A)(B), 5.3.9, and 5.3.9(A)
Delmar, *Firefighter's Handbook*, 2nd Edition, 1st Printing, page 157.
IFSTA, *Essentials of Fire Fighting*, 4th Edition, 1st Printing, page 98.
Jones and Bartlett, *Fundamentals of Fire Fighter Skills*, 1st Edition, 1st Printing, page 58.
Answer: C

27. Reference: NFPA 1001, 5.3.1, 5.3.1(A)(B), 5.3.5, 5.3.5(A)(B), 5.3.9, and 5.3.9(A)
Delmar, *Firefighter's Handbook*, 2nd Edition, 1st Printing, page 158.
IFSTA, *Essentials of Fire Fighting*, 4th Edition, 1st Printing, page 95.
Jones and Bartlett, *Fundamentals of Fire Fighter Skills*, 1st Edition, 1st Printing, page 45.
Answer: D

28. Reference: NFPA 1001, 5.3.1, 5.3.1(A)(B), 5.3.5, 5.3.5(A)(B), 5.3.9, and 5.3.9(A)(B)
Delmar, *Firefighter's Handbook*, 2nd Edition, 1st Printing, page 157.
IFSTA, *Essentials of Fire Fighting*, 4th Edition, 1st Printing, page 98.
Jones and Bartlett, *Fundamentals of Fire Fighter Skills*, 1st Edition, 1st Printing, page 47.
Answer: B

APPENDIX A, EXAMINATION I-3 ANSWER KEY

29. Reference: NFPA 1001, 5.3.1 and 5.3.1(A)(B)
Delmar, *Firefighter's Handbook*, 2nd Edition, 1st Printing, page 143.
IFSTA, *Essentials of Fire Fighting*, 4th Edition, 1st Printing, page 87.
Jones and Bartlett, *Fundamentals of Fire Fighter Skills*, 1st Edition, 1st Printing, pages 42-44.
Answer: B

30. Reference: NFPA 1001, 5.3.1 and 5.3.1(A)(B)
Delmar, *Firefighter's Handbook*, 2nd Edition, 1st Printing, page 145.
IFSTA, *Essentials of Fire Fighting*, 4th Edition, 1st Printing, page 88.
Jones and Bartlett, *Fundamentals of Fire Fighter Skills*, 1st Edition, 1st Printing, page 43.
Answer: B

31. Reference: NFPA 1001, 5.3.1 and 5.3.1(A)(B)
Delmar, *Firefighter's Handbook*, 2nd Edition, 1st Printing, pages 146-148.
IFSTA, *Essentials of Fire Fighting*, 4th Edition, 1st Printing, pages 88-90.
Jones and Bartlett, *Fundamentals of Fire Fighter Skills*, 1st Edition, 1st Printing, pages 42-44.
Answer: D

32. Reference: NFPA 1001, 5.3.1 and 5.3.1(A)(B)
Delmar, *Firefighter's Handbook*, 1st Edition, 1st Printing, page 150.
IFSTA, *Essentials of Fire Fighting*, 4th Edition, 1st Printing, page 94.
Jones and Bartlett, *Fundamentals of Fire Fighter Skills*, 1st Edition, 1st Printing, pages 46-47.
Answer: B

33. Reference: NFPA 1001, 5.3.1 and 5.3.1(A)(B)
Delmar, *Firefighter's Handbook*, 2nd Edition, 1st Printing, page 150.
IFSTA, *Essentials of Fire Fighting*, 4th Edition, 1st Printing, pages 93-94.
Jones and Bartlett, *Fundamentals of Fire Fighter Skills*, 1st Edition, 1st Printing, page 46.
Answer: D

34. Reference: NFPA 1001, 5.3.1 and 5.3.1(A)(B)
Delmar, *Firefighter's Handbook*, 2nd Edition, 1st Printing, page 151.
IFSTA, *Essentials of Fire Fighting*, 4th Edition, 1st Printing, page 95.
Jones and Bartlett, *Fundamentals of Fire Fighter Skills*, 1st Edition, 1st Printing, page 45.
Answer: B

35. Reference: NFPA 1001, 5.3.1, 5.3.1(A)(B), 5.3.5 and 5.3.5(A)(B)
Delmar, *Firefighter's Handbook*, 2nd Edition, 1st Printing, page 171.
IFSTA, *Essentials of Fire Fighting*, 4th Edition, 1st Printing, page 108.
Jones and Bartlett, *Fundamentals of Fire Fighter Skills*, 1st Edition, 1st Printing, page 58.
Answer: C

36. Reference: NFPA 1001, 5.3.1 and 5.3.1(A)(B)
Delmar, *Firefighter's Handbook*, 2nd Edition, 1st Printing, page 170.
IFSTA, *Essentials of Fire Fighting*, 4th Edition, 1st Printing, pages 109-110.
Jones and Bartlett, *Fundamentals of Fire Fighter Skills*, 1st Edition, 1st Printing, page 58.
Answer: B

37. Reference: NFPA 1001, 5.3.2 and 5.3.2(A)(B)
Delmar, *Firefighter's Handbook*, 2nd Edition, 1st Printing, page 120.
IFSTA *Essentials of Fire Fighting*, 4th Edition, 1st Printing, page 23.
Jones and Bartlett, *Fundamentals of Fire Fighter Skills*, 1st Edition, 1st Printing, pages 26 and 281.
Answer: B

38. Reference: NFPA 1001, 5.3.2 and 5.3.2(A)
Delmar, *Firefighter's Handbook*, 2nd Edition, 1st Printing, page 128.
IFSTA, *Essentials of Fire Fighting*, 4th Edition, 1st Printing, page 83.
Jones and Bartlett, *Fundamentals of Fire Fighter Skills*, 1st Edition, 1st Printing, page 32.
Answer: B

39. Reference: NFPA 1001, 5.3.2, 5.3.2(A), 5.3.3 and 5.3.3(A)(B)
Delmar, *Firefighter's Handbook*, 2nd Edition, 1st Printing, page 128.
IFSTA, *Essentials of Fire Fighting*, 4th Edition, 1st Printing, page 83.
Jones and Bartlett, *Fundamentals of Fire Fighter Skills*, 1st Edition, 1st Printing, pages 33-34.
Answer: A

40. Reference: NFPA 1001, 5.3.2 and 5.3.2(A)(B)
Delmar, *Firefighter's Handbook*, 2nd Edition, 1st Printing, page 136.
IFSTA, *Essentials of Fire Fighting*, 4th Edition, 1st Printing, page 86.
Jones and Bartlett, *Fundamentals of Fire Fighter Skills*, 1st Edition, 1st Printing, page 38.
Answer: C

41. Reference: NFPA 1001, 5.3.2 and 5.3.2(A)(B)
Delmar, *Firefighter's Handbook*, 2nd Edition, 1st Printing, pages 130 and 134.
IFSTA, *Essentials of Fire Fighting*, 4th Edition, 1st Printing, page 81.
Jones and Bartlett, *Fundamentals of Fire Fighter Skills*, 1st Edition, 1st Printing, page 37.
Answer: D

42. Reference: NFPA 1001, 5.3.2 and 5.3.2(A)(B)
Delmar, *Firefighter's Handbook*, 2nd Edition, 1st Printing, page 135.
IFSTA, *Essentials of Fire Fighting*, 4th Edition, 1st Printing, pages 80 and 100.
Jones and Bartlett, *Fundamentals of Fire Fighter Skills*, 1st Edition, 1st Printing, page 37.
Answer: B

43. Reference: NFPA 1001, 5.3.4 and 5.3.4(A)(B)
Delmar, *Firefighter's Handbook*, 2nd Edition, 1st Printing, page 523.
IFSTA, *Essentials of Fire Fighting*, 4th Edition, 1st Printing, page 255.
Jones and Bartlett, *Fundamentals of Fire Fighter Skills*, 1st Edition, 1st Printing, page 312.
Answer: B

44. Reference: NFPA 5.3.4 and 5.3.4(A)
Delmar, *Firefighter's Handbook*, 2nd Edition, 1st Printing, page 524.
IFSTA, *Essentials of Fire Fighting*, 4th Edition, 1st Printing, page 246.
Jones and Bartlett, *Fundamentals of Fire Fighter Skills*, 1st Edition, 1st Printing, page 307.
Answer: C

45. Reference: NFPA 5.3.4 and 5.3.4(A)(B)
Delmar, *Firefighter's Handbook*, 2nd Edition, 1st Printing, page 511.
IFSTA, *Essentials of Fire Fighting*, 4th Edition, 1st Printing, page 243.
Jones and Bartlett, *Fundamentals of Fire Fighter Skills*, 1st Edition, 1st Printing, page 299.
Answer: B

46. Reference: NFPA 5.3.4 and 5.3.4(A)(B)
Delmar, *Firefighter's Handbook*, 2nd Edition, 1st Printing, page 519.
IFSTA, *Essentials of Fire Fighting*, 4th Edition, 1st Printing, page 255-256.
Jones and Bartlett, *Fundamentals of Fire Fighter Skills*, 1st Edition, 1st Printing, pages 230 and 303.
Answer: B

47. Reference: NFPA 5.3.4 and 5.3.4(A)(B)
Delmar, *Firefighter's Handbook*, 2nd Edition, 1st Printing, page 525.
IFSTA, *Essentials of Fire Fighting*, 4th Edition, 1st Printing, page 246.
Jones and Bartlett, *Fundamentals of Fire Fighter Skills*, 1st Edition, 1st Printing, page 309.
Answer: D

48. Reference: NFPA 5.3.4 and 5.3.4(A)(B)
Delmar, *Firefighter's Handbook*, 2nd Edition, 1st Printing, page 524, Fig. 17-26.
IFSTA, *Essentials of Fire Fighting*, 4th Edition, 1st Printing, page 243.
Jones and Bartlett, *Fundamentals of Fire Fighter Skills*, 1st Edition, 1st Printing, pages 305-307.
Answer: B

49. Reference: NFPA 5.3.4 and 5.3.4(A)(B)
Delmar, *Firefighter's Handbook*, 2nd Edition, 1st Printing, page 543.
IFSTA, *Essentials of Fire Fighting*, 4th Edition, 1st Printing, page 276.
Jones and Bartlett, *Fundamentals of Fire Fighter Skills*, 1st Edition, 1st Printing, page 312.
Answer: B

50. Reference: NFPA 5.3.4 and 5.3.4(A)(B)
Delmar, *Firefighter's Handbook*, 2nd Edition, 1st Printing, page 543.
IFSTA, *Essentials of Fire Fighting*, 4th Edition, 1st Printing, page 261.
Jones and Bartlett, *Fundamentals of Fire Fighter Skills*, 1st Edition, 1st Printing, page 315.
Answer: D

51. Reference: NFPA 5.3.4 and 5.3.4(A)(B)
Delmar, *Firefighter's Handbook*, 2nd Edition, 1st Printing, page 525.
IFSTA, *Essentials of Fire Fighting*, 4th Edition, 1st Printing, page 248.
Jones and Bartlett, *Fundamentals of Fire Fighter Skills*, 1st Edition, 1st Printing, page 311.
Answer: B

52. Reference: NFPA 5.3.4 and 5.3.4(A)(B)
Delmar, *Firefighter's Handbook*, 2nd Edition, 1st Printing, page 520.
IFSTA, *Essentials of Fire Fighting*, 4th Edition, 1st Printing, page 240.
Jones and Bartlett, *Fundamentals of Fire Fighter Skills*, 1st Edition, 1st Printing, page 299.
Answer: D

53. Reference: NFPA 5.3.4 and 5.3.4(A)
Delmar, *Firefighter's Handbook*, 2nd Edition, 1st Printing, page 510.
IFSTA, *Essentials of Fire Fighting*, 4th Edition, 1st Printing, page 233.
Jones and Bartlett, *Fundamentals of Fire Fighter Skills*, 1st Edition, 1st Printing, page 298.
Answer: C

54. Reference: NFPA 5.3.4 and 5.3.4(A)
Delmar, *Firefighter's Handbook*, 2nd Edition, 1st Printing, pages 515-517.
IFSTA, *Essentials of Fire Fighting*, 4th Edition, 1st Printing, page 234.
Jones and Bartlett, *Fundamentals of Fire Fighter Skills*, 1st Edition, 1st Printing, pages 219 and 226.
Answer: C

55. Reference: NFPA 5.3.4 and 5.3.4(A)
Delmar, *Firefighter's Handbook*, 2nd Edition, 1st Printing, pages 515-516.
Answer: D

56. Reference: NFPA 5.3.4 and 5.3.4(A)
Delmar, *Firefighter's Handbook*, 2nd Edition, 1st Printing, page 517.
IFSTA, *Essentials of Fire Fighting*, 4th Edition, 1st Printing, page 236.
Jones and Bartlett, *Fundamentals of Fire Fighter Skills*, 1st Edition, 1st Printing, page 227.
Answer: D

57. Reference: NFPA 5.3.4 and 5.3.4(A)(B)
Delmar, *Firefighter's Handbook*, 2nd Edition, 1st Printing, page 520.
IFSTA, *Essentials of Fire Fighting*, 4th Edition, 1st Printing, page 357.
Jones and Bartlett, *Fundamentals of Fire Fighter Skills*, 1st Edition, 1st Printing, page 431.
Answer: B

58. Reference: NFPA 5.3.4 and 5.3.4(A)
Delmar, *Firefighter's Handbook*, 2nd Edition, 1st Printing, page 542.
IFSTA, *Essentials of Fire Fighting*, 4th Edition, 1st Printing, page 202.
Jones and Bartlett, *Fundamentals of Fire Fighter Skills*, 1st Edition, 1st Printing, page 312.
Answer: B

59. Reference: NFPA 1001, 5.3.5, 5.3.5(A)(B), 5.3.9, and 5.3.9(A)(B)
Delmar, *Firefighter's Handbook*, 2nd Edition, 1st Printing, page 461.
IFSTA, *Essentials of Fire Fighting*, 4th Edition, 1st Printing, page 183.
Jones and Bartlett, *Fundamentals of Fire Fighter Skills*, 1st Edition, 1st Printing, pages 382 and 540.
Answer: B

60. Reference: NFPA 1001, 5.3.5, 5.3.5(A)(B), 5.3.9, and 5.3.9(A)(B)
Delmar, *Firefighter's Handbook*, 2nd Edition, 1st Printing, page 462.
IFSTA, *Essentials of Fire Fighting*, 4th Edition, 1st Printing, page 183.
Jones and Bartlett, *Fundamentals of Fire Fighter Skills*, 1st Edition, 1st Printing, page 540.
Answer: C

61. Reference: NFPA 1001, 5.3.5, 5.3.5(A)(B), 5.3.9, and 5.3.9(A)(B)
Delmar, *Firefighter's Handbook*, 2nd Edition, 1st Printing, pages 461-462.
IFSTA, *Essentials of Fire Fighting*, 4th Edition, 1st Printing, page 180.
Jones and Bartlett, *Fundamentals of Fire Fighter Skills*, 1st Edition, 1st Printing, page 382.
Answer: D

62. Reference: NFPA 1001, 5.3.5 and 5.3.5(A)(B)
Delmar, *Firefighter's Handbook*, 2nd Edition, 1st Printing, page 736.
IFSTA, *Essentials of Fire Fighting*, 4th Edition, 1st Printing, page 181.
Jones and Bartlett, *Fundamentals of Fire Fighter Skills*, 1st Edition, 1st Printing, page 545.
Answer: D

63. Reference: NFPA 1001, 5.3.5 and 5.3.5(A)(B)
Delmar, *Firefighter's Handbook*, 2nd Edition, 1st Printing, page 725.
IFSTA, *Essentials of Fire Fighting*, 4th Edition, 1st Printing, pages 29-30.
Jones and Bartlett, *Fundamentals of Fire Fighter Skills*, 1st Edition, 1st Printing, pages 285 and 540.
Answer: B

64. References: NFPA 1001, 5.3.6 and 5.3.6(A)(B)
Delmar, *Firefighter's Handbook*, 2nd Edition, 1st Printing, pages 391-392 and 399.
IFSTA, *Essentials of Fire Fighting*, 4th Edition, 1st Printing, page 298.
Jones and Bartlett, *Fundamentals of Fire Fighter Skills*, 1st Edition, 1st Printing, pages 340-341.
Answer: B

65. References: NFPA 1001, 5.3.6, 5.3.6(A)(B), 5.3.12 and 5.3.12(A)(B)
Delmar, *Firefighter's Handbook*, 2nd Edition, 1st Printing, page 395.
IFSTA, *Essentials of Fire Fighting*, 4th Edition, 1st Printing, page 311.
Jones and Bartlett, *Fundamentals of Fire Fighter Skills*, 1st Edition, 1st Printing, pages 348-349.
Answer: D

66. References: NFPA 1001, 5.3.6, 5.3.6(A)(B), 5.3.12 and 5.3.12(A)(B)
Delmar, *Firefighter's Handbook*, 2nd Edition, 1st Printing, page 402.
IFSTA, *Essentials of Fire Fighting*, 4th Edition, 1st Printing, page 300.
Jones and Bartlett, *Fundamentals of Fire Fighter Skills*, 1st Edition, 1st Printing, pages 353-354.
Answer: B

67. References: NFPA 1001, 5.3.6, 5.3.6(A)(B), 5.3.11, 5.3.11(A)(B), 5.3.12, and 5.3.12(A)(B)
Delmar, *Firefighter's Handbook*, 2nd Edition, 1st Printing, pages 384-385.
IFSTA, *Essentials of Fire Fighting*, 4th Edition, 1st Printing, page 291.
Jones and Bartlett, *Fundamentals of Fire Fighter Skills*, 1st Edition, 1st Printing, pages 342 and 352.
Answer: C

68. References: NFPA 1001, 5.3.6, 5.3.6(A)(B), 5.3.11, 5.3.11(A)(B), 5.3.11(A)(B), 5.3.12 and 5.3.12(A)(B)
Delmar, *Firefighter's Handbook*, 2nd Edition, 1st Printing, page 386.
IFSTA, *Essentials of Fire Fighting*, 4th Edition, 1st Printing, page 296.
Jones and Bartlett, *Fundamentals of Fire Fighter Skills*, 1st Edition, 1st Printing, pages 342-343.
Answer: B

69. References: NFPA 1001, 5.3.6 and 5.3.6(A)(B)
Delmar, *Firefighter's Handbook*, 2nd Edition, 1st Printing, pages 392 and 408.
IFSTA, *Essentials of Fire Fighting*, 4th Edition, 1st Printing, pages 304, and 340-341.
Jones and Bartlett, *Fundamentals of Fire Fighter Skills*, 1st Edition, 1st Printing, pages 364-365.
Answer: D

70. References: NFPA 1001, 5.3.6, 5.3.6(A)(B), 5.3.11, 5.3.11(A)(B), 5.3.12 and 5.3.12(A)(B)
Delmar, *Firefighter's Handbook*, 2nd Edition, 1st Printing, pages 384-385.
IFSTA, *Essentials of Fire Fighting*, 4th Edition, 1st Printing, page 291.
Jones and Bartlett, *Fundamentals of Fire Fighter Skills*, 1st Edition, 1st Printing, pages 342 and 352.
Answer: D

71. References: NFPA 1001, 5.3.6, 5.3.6(A)(B), 5.3.11, 5.3.11(A)(B), 5.3.12 and 5.3.12(A)(B)
Delmar, *Firefighter's Handbook*, 2nd Edition, 1st Printing, pages 384-385.
IFSTA, *Essentials of Fire Fighting*, 4th Edition, 1st Printing, page 291.
Jones and Bartlett, *Fundamentals of Fire Fighter Skills*, 1st Edition, 1st Printing, pages 342 and 352.
Answer: C

72. References: NFPA 1001, 5.3.6 and 5.3.6(A)
Delmar, *Firefighter's Handbook*, 1st Edition, 2nd Printing, page 371.
IFSTA, *Essentials of Fire Fighting*, 4th Edition, 1st Printing, page 281.
Jones and Bartlett, *Fundamentals of Fire Fighter Skills*, 1st Edition, 1st Printing, pages 332-333.
Answer: B

73. References: NFPA 1001, 5.3.6 and 5.3.6(A)(B)
Delmar, *Firefighter's Handbook*, 2nd Edition, 1st Printing, pages 400-403.
IFSTA, *Essentials of Fire Fighting*, 4th Edition, 1st Printing, page 301.
Jones and Bartlett, *Fundamentals of Fire Fighter Skills*, 1st Edition, 1st Printing, page 352.
Answer: B

74. References: NFPA 1001, 5.3.6, 5.3.6(A)(B), 5.3.12, and 5.3.12(A)(B)
Delmar, *Firefighter's Handbook*, 2nd Edition, 1st Printing, page 383.
IFSTA, *Essentials of Fire Fighting*, 4th Edition, 1st Printing, page 305.
Jones and Bartlett, *Fundamentals of Fire Fighter Skills*, 1st Edition, 1st Printing, page 364.
Answer: A

75. References: NFPA 1001, 5.3.6 and 5.3.6(A)(B)
Delmar, *Firefighter's Handbook*, 1st Edition, 2nd Printing, page 412.
IFSTA, *Essentials of Fire Fighting*, 4th Edition, 1st Printing, pages 327 and 330.
Jones and Bartlett, *Fundamentals of Fire Fighter Skills*, 1st Edition, 1st Printing, pages 366-367.
Answer: C

76. References: NFPA 1001, 5.3.6 and 5.3.6(A)
Delmar, *Firefighter's Handbook*, 2nd Edition, 1st Printing, page 372.
IFSTA, *Essentials of Fire Fighting*, 4th Edition, 1st Printing, pages 288-289.
Jones and Bartlett, *Fundamentals of Fire Fighter Skills*, 1st Edition, 1st Printing, page 332.
Answer: B

77. References: NFPA 1001, 5.3.6 and 5.3.6(A)
Delmar, *Firefighter's Handbook*, 2nd Edition, 1st Printing, page 373.
IFSTA, *Essentials of Fire Fighting*, 4th Edition, 1st Printing, page 543.
Jones and Bartlett, *Fundamentals of Fire Fighter Skills*, 1st Edition, 1st Printing, page 9.
Answer: A

78. References: NFPA 1001, 5.3.6 and 5.3.6(A)
Delmar, *Firefighter's Handbook*, 1st Edition, 1st Printing, page 377.
IFSTA, *Essentials of Fire Fighting*, 4th Edition, 1st Printing, page 286.
Jones and Bartlett, *Fundamentals of Fire Fighter Skills*, 1st Edition, 1st Printing, page 336.
Answer: C

79. References: NFPA 1001, 5.3.6 and 5.3.6(A)(B)
Delmar, *Firefighter's Handbook*, 2nd Edition, 1st Printing, page 400.
IFSTA, *Essentials of Fire Fighting*, 4th Edition, 1st Printing, page 298.
Jones and Bartlett, *Fundamentals of Fire Fighter Skills*, 1st Edition, 1st Printing, page 352.
Answer: B

80. Reference: NFPA 1001, 5.3.7 and 5.3.7(A)(B)
Delmar, *Firefighter's Handbook*, 2nd Edition, 1st Printing, page 623.
IFSTA, *Essentials of Fire Fighting*, 4th Edition, 1st Printing, page 548.
Jones and Bartlett, *Fundamentals of Fire Fighter Skills*, 1st Edition, 1st Printing, page 637.
Answer: A

81. Reference: NFPA 5.3.7 and 5.3.7(A)(B)
Delmar, *Firefighter's Handbook*, 2nd Edition, 1st Printing, page 600.
IFSTA, *Essentials of Fire Fighting*, 4th Edition, 1st Printing, page 549.
Jones and Bartlett, *Fundamentals of Fire Fighter Skills*, 1st Edition, 1st Printing, page 637.
Answer: D

82. Reference: NFPA 1001, 5.3.9 and 5.3.9(A)(B)
Delmar, *Firefighter's Handbook*, 2nd Edition, 1st Printing, page 465.
IFSTA, *Essentials of Fire Fighting*, 4th Edition, 1st Printing, page 177.
Jones and Bartlett, *Fundamentals of Fire Fighter Skills*, 1st Edition, 1st Printing, page 378.
Answer: C

83. Reference: NFPA 1001, 5.3.9 and 5.3.9(A)(B)
Delmar, *Firefighter's Handbook*, 2nd Edition, 1st Printing, page 465.
IFSTA, *Essentials of Fire Fighting*, 4th Edition, 1st Printing, pages 176-177.
Jones and Bartlett, *Fundamentals of Fire Fighter Skills*, 1st Edition, 1st Printing, pages 378-379.
Answer: B

84. Reference: NFPA 1001, 5.3.9 and 5.3.9(A)
Delmar, *Firefighter's Handbook*, 2nd Edition, 1st Printing, page 461.
IFSTA, *Essentials of Fire Fighting*, 4th Edition, 1st Printing, page 180.
Jones and Bartlett, *Fundamentals of Fire Fighter Skills*, 1st Edition, 1st Printing, page 375.
Answer: B

85. Reference: NFPA 1001, 5.3.9 and 5.3.9(A)(B)
Delmar, *Firefighter's Handbook*, 2nd Edition, 1st Printing, page 463.
IFSTA, *Essentials of Fire Fighting*, 4th Edition, 1st Printing, page 179.
Jones and Bartlett, *Fundamentals of Fire Fighter Skills*, 1st Edition, 1st Printing, pages 379-380.
Answer: D

86. Reference: NFPA 1001, 5.3.9 and 5.3.9(A)(B)
Delmar, *Firefighter's Handbook*, 2nd Edition, 1st Printing, page 465.
IFSTA, *Essentials of Fire Fighting*, 4th Edition, 1st Printing, pages 176-178.
Jones and Bartlett, *Fundamentals of Fire Fighter Skills*, 1st Edition, 1st Printing, pages 378-382.
Answer: A

87. Reference: NFPA 1001, 5.3.9 and 5.3.9(A)(B)
Delmar, *Firefighter's Handbook*, 2nd Edition, 1st Printing, page 467.
IFSTA, *Essentials of Fire Fighting*, 4th Edition, 1st Printing, page 184.
Jones and Bartlett, *Fundamentals of Fire Fighter Skills*, 1st Edition, 1st Printing, page 386.
Answer: B

88. Reference: NFPA 1001, 5.3.9 and 5.3.9(A)(B)
Delmar, *Firefighter's Handbook*, 2nd Edition, 1st Printing, page 467.
IFSTA, *Essentials of Fire Fighting*, 4th Edition, 1st Printing, pages 183-184.
Jones and Bartlett, *Fundamentals of Fire Fighter Skills*, 1st Edition, 1st Printing, pages 391-395.
Answer: B

89. Reference: NFPA 1001, 5.3.10 and 5.3.10(A)
Delmar, *Firefighter's Handbook*, 2nd Edition, 1st Printing, page 364.
IFSTA, *Essentials of Fire Fighting*, 4th Edition, 1st Printing, pages 73-74.
Jones and Bartlett, *Fundamentals of Fire Fighter Skills*, 1st Edition, 1st Printing, pages 383-384 and 628.
Answer: D

90. Reference: NFPA 1001, 5.3.10 and 5.3.10(A)(B)
Delmar, *Firefighter's Handbook*, 2nd Edition, 1st Printing, page 283.
IFSTA, *Essentials of Fire Fighting*, 4th Edition, 1st Printing, page 493.
Jones and Bartlett, *Fundamentals of Fire Fighter Skills*, 1st Edition, 1st Printing, page 520.
Answer: A

91. Reference: NFPA 1001, 5.3.10, 5.3.10(A)(B), 5.5.4, and 5.5.4(A)(B)
Delmar, *Firefighter's Handbook*, 2nd Edition, 1st Printing, page 231.
IFSTA, *Essentials of Fire Fighting*, 4th Edition, 1st Printing, pages 415 and 445.
Jones and Bartlett, *Fundamentals of Fire Fighter Skills*, 1st Edition, 1st Printing, pages 484 and 488.
Answer: C

92. Reference: NFPA 1001, 5.3.10 and 5.3.10(A)(B)
Delmar, *Firefighter's Handbook*, 2nd Edition, 1st Printing, page 253.
IFSTA, *Essentials of Fire Fighting*, 4th Edition, 1st Printing, pages 430 and 522.
Jones and Bartlett, *Fundamentals of Fire Fighter Skills*, 1st Edition, 1st Printing, page 514.
Answer: A

93. Reference: NFPA 1001, 5.3.10 and 5.3.10(A)(B)
Delmar, *Firefighter's Handbook*, 1st Edition, 1st Printing, page 217.
IFSTA, *Essentials of Fire Fighting*, 4th Edition, 1st Printing, page 491.
Jones and Bartlett, *Fundamentals of Fire Fighter Skills*, 1st Edition, 1st Printing, pages 467 and 533.
Answer: B

94. Reference: NFPA 1001, 5.3.10 and 5.3.10(A)(B)
Delmar, *Firefighter's Handbook*, 2nd Edition, 1st Printing, page 284.
IFSTA, *Essentials of Fire Fighting*, 4th Edition, 1st Printing, page 527.
Jones and Bartlett, *Fundamentals of Fire Fighter Skills*, 1st Edition, 1st Printing, page 627.
Answer: D

95. Reference: NFPA 1001, 5.3.10, 5.3.10(A)(B), 5.3.8 and 5.3.8(A)(B)
Delmar, *Firefighter's Handbook*, 2nd Edition, 1st Printing, pages 283-284.
IFSTA, *Essentials of Fire Fighting*, 4th Edition, 1st Printing, page 494.
Jones and Bartlett, *Fundamentals of Fire Fighter Skills*, 1st Edition, 1st Printing, page 626.
Answer: B

96. Reference: NFPA 1001, 5.3.10, 5.3.10(A)(B), 5.3.8 and 5.3.8(A)(B)
Delmar, *Firefighter's Handbook*, 2nd Edition, 1st Printing, pages 271 and 289.
IFSTA, *Essentials of Fire Fighting*, 4th Edition, 1st Printing, page 492.
Jones and Bartlett, *Fundamentals of Fire Fighter Skills*, 1st Edition, 1st Printing, page 520.
Answer: C

97. Reference: NFPA 1001, 5.3.10, 5.3.10(A), 5.3.8 and 5.3.8(A)(B)
Delmar, *Firefighter's Handbook*, 2nd Edition, 1st Printing, pages 282-283.
IFSTA, *Essentials of Fire Fighting*, 4th Edition, 1st Printing, pages 492-496.
Jones and Bartlett, *Fundamentals of Fire Fighter Skills*, 1st Edition, 1st Printing, page 521.
Answer: A

98. Reference: NFPA 1001, 5.3.10, 5.3.10(A)(B), 5.3.8, and 5.3.8(A)(B)
Delmar, *Firefighter's Handbook*, 2nd Edition, 1st Printing, page 603.
IFSTA, *Essentials of Fire Fighting*, 4th Edition, 1st Printing, page 525.
Jones and Bartlett, *Fundamentals of Fire Fighter Skills*, 1st Edition, 1st Printing, page 635.
Answer: D

99. Reference: NFPA 1001, 5.3.10, 5.3.10(A)(B), 5.3.8 and 5.3.8(A)(B)
Delmar, *Firefighter's Handbook*, 2nd Edition, 1st Printing, page 604.
IFSTA, *Essentials of Fire Fighting*, 4th Edition, 1st Printing, page 527.
Jones and Bartlett, Fundamentals of Firefighter Skills, 1st Edition, 1st Printing, page 628.
Answer: C

100. Reference: NFPA 1001, 5.3.10, 5.3.10(A)(B), 5.3.8 and 5.3.8(A)(B)
Delmar, *Firefighter's Handbook*, 1st Edition, 2nd Printing, pages 283-284.
IFSTA, *Essentials of Fire Fighting*, 4th Edition, 1st Printing, page 496.
Jones and Bartlett, *Fundamentals of Fire Fighter Skills*, 1st Edition, 1st Printing, page 523.
Answer: D

101. Reference: NFPA 1001, 5.3.10, 5.3.10(A)(B), 5.3.8 and 5.3.8(A)(B)
Delmar, *Firefighter's Handbook*, 2nd Edition, 1st Printing, pages 257-258.
IFSTA, *Essentials of Fire Fighting*, 4th Edition, 1st Printing, page 432.
Jones and Bartlett, *Fundamentals of Fire Fighter Skills*, 1st Edition, 1st Printing, page 516.
Answer: C

102. Reference: NFPA 1001, 5.3.10 and 5.3.10(A)(B)
Delmar, *Firefighter's Handbook*, 2nd Edition, 1st Printing, page 253.
IFSTA, *Essentials of Fire Fighting*, 4th Edition, 1st Printing, page 430.
Jones and Bartlett, *Fundamentals of Fire Fighter Skills*, 1st Edition, 1st Printing, page 514.
Answer: B

103. Reference: NFPA 1001, 5.3.10, 5.3.10(A)(B), 5.3.8 and 5.3.8(A)(B)
Delmar, *Firefighter's Handbook*, 2nd Edition, 1st Printing, page 227.
IFSTA, *Essentials of Fire Fighting*, 4th Edition, 1st Printing, page 412.
Jones and Bartlett, *Fundamentals of Fire Fighter Skills*, 1st Edition, 1st Printing, pages 469-470.
Answer: D

104. Reference: NFPA 1001, 5.3.10, 5.3.10(A)(B), 5.3.8 and 5.3.8(A)(B)
Delmar, *Firefighter's Handbook*, 2nd Edition, 1st Printing, page 227.
IFSTA, *Essentials of Fire Fighting*, 4th Edition, 1st Printing, pages 410-411.
Jones and Bartlett, *Fundamentals of Fire Fighter Skills*, 1st Edition, 1st Printing, page 482.
Answer: C

105. Reference: NFPA 1001, 5.3.10 and 5.3.10(A)(B)
Delmar, *Firefighter's Handbook*, 2nd Edition, 1st Printing, pages 255-257 and 616-618.
IFSTA, *Essentials of Fire Fighting*, 4th Edition, 1st Printing, page 431.
Jones and Bartlett, *Fundamentals of Fire Fighter Skills*, 1st Edition, 1st Printing, page 518.
Answer: D

106. Reference: NFPA 1001, 5.3.10, 5.3.10(A)(B), 5.3.8 and 5.3.8(A)(B)
Delmar, *Firefighter's Handbook*, 2nd Edition, 1st Printing, pages 230-231.
IFSTA, *Essentials of Fire Fighting*, 4th Edition, 1st Printing, page 448.
Jones and Bartlett, *Fundamentals of Fire Fighter Skills*, 1st Edition, 1st Printing, pages 474-475.
Answer: A

107. Reference: NFPA 1001, 5.3.10, 5.3.10(A), 5.3.8 and 5.3.8(A)(B)
Delmar, *Firefighter's Handbook*, 2nd Edition, 1st Printing, pages 216-217.
IFSTA, *Essentials of Fire Fighting*, 4th Edition, 1st Printing, page 491.
Jones and Bartlett, *Fundamentals of Fire Fighter Skills*, 1st Edition, 1st Printing, pages 467 and 533.
Answer: D

108. Reference: NFPA 1001, 5.3.10, 5.3.10(A)(B), 5.3.8 and 5.3.8(A)(B)
Delmar, *Firefighter's Handbook*, 2nd Edition, 1st Printing, page 237.
IFSTA, *Essentials of Fire Fighting*, 4th Edition, 1st Printing, page 416.
Jones and Bartlett, *Fundamentals of Fire Fighter Skills*, 1st Edition, 1st Printing, pages 494 and 532.
Answer: A

109. Reference: NFPA 1001, 5.3.10, 5.3.10(A)(B), 5.3.8 and 5.3.8(A)(B)
Delmar, *Firefighter's Handbook*, 2nd Edition, 1st Printing, page 237.
IFSTA, *Essentials of Fire Fighting*, 4th Edition, 1st Printing, page 417.
Jones and Bartlett, *Fundamentals of Fire Fighter Skills*, 1st Edition, 1st Printing, page 497.
Answer: D

110. Reference: NFPA 1001, 5.3.10, 5.3.10(A)(B), 5.3.7, 5.3.7(A), 5.3.8 and 5.3.8(A)
Delmar, *Firefighter's Handbook*, 2nd Edition, 1st Printing, page 289.
IFSTA, *Essentials of Fire Fighting*, 4th Edition, 1st Printing, pages 525-527.
Jones and Bartlett, *Fundamentals of Fire Fighter Skills*, 1st Edition, 1st Printing, page 627.
Answer: A

111. Reference: NFPA 1001, 5.3.11, 5.3.11(A) 5.3.12, and 5.3.12(A)
Delmar, *Firefighter's Handbook*, 2nd Edition, 1st Printing, pages 588-589.
IFSTA, *Essentials of Fire Fighting*, 4th Edition, 1st Printing, page 364.
Jones and Bartlett, *Fundamentals of Fire Fighter Skills*, 1st Edition, 1st Printing, page 415.
Answer: B

112. Reference: NFPA 1001, 5.3.11, 5.3.11(A) 5.3.12, and 5.3.12(A)
Delmar, *Firefighter's Handbook*, 2nd Edition, 1st Printing, page 554.
IFSTA, *Essentials of Fire Fighting*, 4th Edition, 1st Printing, page 364.
Jones and Bartlett, *Fundamentals of Fire Fighter Skills*, 1st Edition, 1st Printing, pages 411 and 415.
Answer: A

113. Reference: NFPA 1001, 5.3.11, 5.3.11(A)(B), 5.3.12, and 5.3.12(A)(B)
Delmar, *Firefighter's Handbook*, 2nd Edition, 1st Printing, pages 567-574.
IFSTA, *Essentials of Fire Fighting*, 4th Edition, 1st Printing, pages 345-346.
Jones and Bartlett, *Fundamentals of Fire Fighter Skills*, 1st Edition, 1st Printing, pages 414, 419, and 423.
Answer: D

114. Reference: NFPA 1001, 5.3.11 and 5.3.11(A)(B)
Delmar, *Firefighter's Handbook*, 2nd Edition, 1st Printing, page 563.
IFSTA, *Essentials of Fire Fighting*, 4th Edition, 1st Printing, page 366.
Jones and Bartlett, *Fundamentals of Fire Fighter Skills*, 1st Edition, 1st Printing, page 411.
Answer: B

115. Reference: NFPA 1001, 5.3.11 and 5.3.11(A)(B)
Delmar, *Firefighter's Handbook*, 2nd Edition, 1st Printing, pages 563-565.
IFSTA, *Essentials of Fire Fighting*, 4th Edition, 1st Printing, pages 369-370.
Jones and Bartlett, *Fundamentals of Fire Fighter Skills*, 1st Edition, 1st Printing, page 411.
Answer: C

116. Reference: NFPA 1001, 5.3.11, 5.3.11(A), 5.3.12 and 5.3.12(A)
Delmar, *Firefighter's Handbook*, 2nd Edition, 1st Printing, pages 92-93.
IFSTA, *Essentials of Fire Fighting*, 4th Edition, 1st Printing, pages 51-52.
Jones and Bartlett, *Fundamentals of Fire Fighter Skills*, 1st Edition, 1st Printing, page 137.
Answer: D

117. Reference: NFPA 1001, 5.3.11, 5.3.11(A), 5.3.13 and 5.3.13(A)
Delmar, *Firefighter's Handbook*, 2nd Edition, 1st Printing, pages 94-97.
IFSTA, *Essentials of Fire Fighting*, 4th Edition, 1st Printing, page 36.
Jones and Bartlett, *Fundamentals of Fire Fighter Skills*, 1st Edition, 1st Printing, page 130.
Answer: B

118. Reference: NFPA 1001, 5.3.11, 5.3.11(A), 5.3.12, 5.3.12(A), 5.3.13 and 5.3.13(A)
Delmar, *Firefighter's Handbook*, 2nd Edition, 1st Printing, page 100.
IFSTA, *Essentials of Fire Fighting*, 4th Edition, 1st Printing, page 56.
Jones and Bartlett, *Fundamentals of Fire Fighter Skills*, 1st Edition, 1st Printing, page 125.
Answer: B

119. Reference: NFPA 1001, 5.3.11, 5.3.11(A), 5.3.12, 5.3.12(A), 5.3.13 and 5.3.13(A)
Delmar, *Firefighter's Handbook*, 2nd Edition, 1st Printing, page 86.
IFSTA, *Essentials of Fire Fighting*, 4th Edition, 1st Printing, page 42.
Jones and Bartlett, *Fundamentals of Fire Fighter Skills*, 1st Edition, 1st Printing, page 127.
Answer: B

120. Reference: NFPA 1001, 5.3.11, 5.3.11(A), 5.3.12, 5.3.12(A), 5.3.13 and 5.3.13(A)
Delmar, *Firefighter's Handbook*, 2nd Edition, 1st Printing, page 94.
IFSTA, *Essentials of Fire Fighting*, 4th Edition, 1st Printing, page 37.
Jones and Bartlett, *Fundamentals of Fire Fighter Skills*, 1st Edition, 1st Printing, page 130.
Answer: A

121. Reference: NFPA 1001, 5.3.11, 5.3.11(A), 5.3.12, 5.3.12(A), 5.3.13 and 5.3.13(A)
Delmar, *Firefighter's Handbook*, 2nd Edition, 1st Printing, page 87.
IFSTA, *Essentials of Fire Fighting*, 4th Edition, 1st Printing, page 39.
Jones and Bartlett, *Fundamentals of Fire Fighter Skills*, 1st Edition, 1st Printing, page 133.
Answer: B

122. Reference: NFPA 1001, 5.3.11, 5.3.11(A), 5.3.12, 5.3.12(A), 5.3.13 and 5.3.13(A)
Delmar, *Firefighter's Handbook*, 2nd Edition, 1st Printing, pages 145-146.
IFSTA, *Essentials of Fire Fighting*, 4th Edition, 1st Printing, pages 55-56.
Jones and Bartlett, *Fundamentals of Fire Fighter Skills*, 1st Edition, 1st Printing, page 129.
Answer: C

123. Reference: NFPA 1001, 5.3.11, 5.3.11(A), 5.3.12, 5.3.12(A), 5.3.13 and 5.3.13(A)
Delmar, *Firefighter's Handbook*, 2nd Edition, 1st Printing, pages 94-95.
IFSTA, *Essentials of Fire Fighting*, 4th Edition, 1st Printing, pages 36-37.
Jones and Bartlett, *Fundamentals of Fire Fighter Skills*, 1st Edition, 1st Printing, page 131.
Answer: A

124. Reference: NFPA 1001, 5.3.11, 5.3.11(A), 5.3.10, 5.3.10(A), 5.3.12, 5.3.12(A), 5.3.13 and 5.3.13(A)
Delmar, *Firefighter's Handbook*, 2nd Edition, 1st Printing, page 92.
IFSTA, *Essentials of Fire Fighting*, 4th Edition, 1st Printing, page 51.
Jones and Bartlett, *Fundamentals of Fire Fighter Skills*, 1st Edition, 1st Printing, page 137.
Answer: D

125. Reference: NFPA 1001, 5.3.11, 5.3.11(A), 5.3.12, and 5.3.12(A)
Delmar, *Firefighter's Handbook*, 2nd Edition, 1st Printing, pages 80 and 104.
IFSTA, *Essentials of Fire Fighting*, 4th Edition, 1st Printing, page 42.
Jones and Bartlett, *Fundamentals of Fire Fighter Skills*, 1st Edition, 1st Printing, page 805.
Answer: C

126. Reference: NFPA 1001, 5.3.11, 5.3.11(A), 5.3.12, and 5.3.12(A)
Delmar, *Firefighter's Handbook*, 2nd Edition, 1st Printing, pages 87-88.
IFSTA, *Essentials of Fire Fighting*, 4th Edition, 1st Printing, page 39.
Jones and Bartlett, *Fundamentals of Fire Fighter Skills*, 1st Edition, 1st Printing, page 133.
Answer: D

APPENDIX A, EXAMINATION I-3 ANSWER KEY

127. Reference: NFPA 1001, 5.3.11, 5.3.11(A), 5.3.12, and 5.3.12(A)
Delmar, *Firefighter's Handbook*, 2nd Edition, 1st Printing, pages 91 and 104.
IFSTA, *Essentials of Fire Fighting*, 4th Edition, 1st Printing, pages 48-49.
Jones and Bartlett, *Fundamentals of Fire Fighter Skills*, 1st Edition, 1st Printing, page 133.
Answer: B

128. Reference: NFPA 1001, 5.3.11, 5.3.11(A), 5.3.10, 5.3.10(A), 5.3.12, and 5.3.12(A)
Delmar, *Firefighter's Handbook*, 2nd Edition, 1st Printing, page 97.
IFSTA, *Essentials of Fire Fighting*, 4th Edition, 1st Printing, page 49.
Jones and Bartlett, *Fundamentals of Fire Fighter Skills*, 1st Edition, 1st Printing, page 138.
Answer: A

129. Reference: NFPA 1001, 5.3.12 and 5.3.12(A)(B)
Delmar, *Firefighter's Handbook*, 2nd Edition, 1st Printing, page 410.
IFSTA, *Essentials of Fire Fighting*, 4th Edition, 1st Printing, pages 372 and 374.
Jones and Bartlett, *Fundamentals of Fire Fighter Skills*, 1st Edition, 1st Printing, page 424.
Answer: B

130. Reference: NFPA 1001, 5.3.12 and 5.3.12(A)(B)
Delmar, *Firefighter's Handbook*, 2nd Edition, 1st Printing, pages 579-581.
IFSTA, *Essentials of Fire Fighting*, 4th Edition, 1st Printing, page 361.
Jones and Bartlett, *Fundamentals of Fire Fighter Skills*, 1st Edition, 1st Printing, pages 434-436.
Answer: A

131. Reference: NFPA 1001, 5.3.12 and 5.3.12(A)(B)
Delmar, *Firefighter's Handbook*, 2nd Edition, 1st Printing, page 581.
IFSTA, *Essentials of Fire Fighting*, 4th Edition, 1st Printing, page 363.
Jones and Bartlett, *Fundamentals of Fire Fighter Skills*, 1st Edition, 1st Printing, page 424.
Answer: A

132. Reference: NFPA 1001, 5.3.12, 5.3.12(A)(B), 5.3.10 and 5.3.10(A)(B)
Delmar, *Firefighter's Handbook*, 2nd Edition, 1st Printing, page 579.
IFSTA, *Essentials of Fire Fighting*, 4th Edition, 1st Printing, page 75.
Jones and Bartlett, *Fundamentals of Fire Fighter Skills*, 1st Edition, 1st Printing, page 427.
Answer: C

133. Reference: NFPA 1001, 5.3.12 and 5.3.12(A)(B)
Delmar, *Firefighter's Handbook*, 2nd Edition, 1st Printing, page 579.
IFSTA, *Essentials of Fire Fighting*, 4th Edition, 1st Printing, page 361.
Jones and Bartlett, *Fundamentals of Fire Fighter Skills*, 1st Edition, 1st Printing, page 435.
Answer: C

134. Reference: NFPA 1001, 5.3.12 and 5.3.12(A)(B)
Delmar, *Firefighter's Handbook*, 2nd Edition, 1st Printing, pages 550-551.
IFSTA, *Essentials of Fire Fighting*, 4th Edition, 1st Printing, pages 345 and 347.
Jones and Bartlett, *Fundamentals of Fire Fighter Skills*, 1st Edition, 1st Printing, page 410.
Answer: D

135. Reference: NFPA 1001, 5.3.12, 5.3.12(A)(B), 5.3.11 and 5.3.11(A)(B)
Delmar, *Firefighter's Handbook*, 2nd Edition, 1st Printing, page 550.
IFSTA, *Essentials of Fire Fighting*, 4th Edition, 1st Printing, page 345.
Jones and Bartlett, *Fundamentals of Fire Fighter Skills*, 1st Edition, 1st Printing, pages 408 and 410.
Answer: C

136. Reference: NFPA 1001, 5.3.12, 5.3.12(A)(B), 5.3.11 and 5.3.11(A)(B)
Delmar, *Firefighter's Handbook*, 2nd Edition, 1st Printing, page 550.
IFSTA, *Essentials of Fire Fighting*, 4th Edition, 1st Printing, page 345.
Jones and Bartlett, *Fundamentals of Fire Fighter Skills*, 1st Edition, 1st Printing, page 410.
Answer: B

137. Reference: NFPA 1001, 5.3.12, 5.3.12(A)(B), 5.3.11 and 5.3.11(A)(B)
Delmar, *Firefighter's Handbook*, 2nd Edition, 1st Printing, page 563.
IFSTA, *Essentials of Fire Fighting*, 4th Edition, 1st Printing, page 366.
Jones and Bartlett, *Fundamentals of Fire Fighter Skills*, 1st Edition, 1st Printing, pages 414-419.
Answer: A

138. Reference: NFPA 1001, 5.3.12, 5.3.12(A), 5.3.10, 5.3.10(A), 5.3.11 and 5.3.11(A)
Delmar, *Firefighter's Handbook*, 2nd Edition, 1st Printing, pages 356-357 and 366.
IFSTA, *Essentials of Fire Fighting*, 4th Edition, 1st Printing, page 361.
Jones and Bartlett, *Fundamentals of Fire Fighter Skills*, 1st Edition, 1st Printing, page 412.
Answer: C

139. Reference: NFPA 1001, 5.3.12, 5.3.12(A), 5.3.11 and 5.3.11(A)
Delmar, *Firefighter's Handbook*, 2nd Edition, 1st Printing, pages 342 and 366.
Jones and Bartlett, *Fundamentals of Fire Fighter Skills*, 1st Edition, 1st Printing, page 159.
Answer: D

140. Reference: NFPA 1001, 5.3.12, 5.3.12(A), 5.3.11 and 5.3.11(A)
Delmar, *Firefighter's Handbook*, 2nd Edition, 1st Printing, page 349.
IFSTA, *Essentials of Fire Fighting*, 4th Edition, 1st Printing, page 70.
Jones and Bartlett, *Fundamentals of Fire Fighter Skills*, 1st Edition, 1st Printing, page 150.
Answer: B

141. Reference: NFPA 1001, 5.3.13 and 5.3.13(A)(B)
Delmar, *Firefighter's Handbook*, 2nd Edition, 1st Printing, page 649.
IFSTA, *Essentials of Fire Fighting*, 4th Edition, 1st Printing, page 596.
Jones and Bartlett, *Fundamentals of Fire Fighter Skills*, 1st Edition, 1st Printing, page 574.
Answer: A

142. Reference: NFPA 1001, 5.3.13 and 5.3.13(A)(B)
Delmar, *Firefighter's Handbook*, 2nd Edition, 1st Printing, page 649.
IFSTA, *Essentials of Fire Fighting*, 4th Edition, 1st Printing, page 597.
Jones and Bartlett, *Fundamentals of Fire Fighter Skills*, 1st Edition, 1st Printing, page 576.
Answer: A

143. Reference: NFPA 1001, 5.3.13 and 5.3.13(A)(B)
Delmar, *Firefighter's Handbook*, 2nd Edition, 1st Printing, page 649.
IFSTA, *Essentials of Fire Fighting*, 4th Edition, 1st Printing, page 597.
Jones and Bartlett, *Fundamentals of Fire Fighter Skills*, 1st Edition, 1st Printing, page 576.
Answer: C

144. Reference: NFPA 1001, 5.3.13, 5.3.13(A)(B), 5.3.14 and 5.3.14(A)(B)
Delmar, *Firefighter's Handbook*, 2nd Edition, 1st Printing, page 649.
IFSTA, *Essentials of Fire Fighting*, 4th Edition, 1st Printing, page 627.
Jones and Bartlett, *Fundamentals of Fire Fighter Skills*, 1st Edition, 1st Printing, pages 967-974.
Answer: A

145. Reference: NFPA 1001, 5.3.13 and 5.3.13(A)(B)
Delmar, *Firefighter's Handbook*, 2nd Edition, 1st Printing, page 649.
IFSTA, *Essentials of Fire Fighting*, 4th Edition, 1st Printing, page 598.
Jones and Bartlett, *Fundamentals of Fire Fighter Skills*, 1st Edition, 1st Printing, page 576.
Answer: C

146. Reference: NFPA 1001, 5.3.13 and 5.3.13(A)(B)
Delmar, *Firefighter's Handbook*, 2nd Edition, 1st Printing, page 652.
IFSTA, *Essentials of Fire Fighting*, 4th Edition, 1st Printing, page 622.
Jones and Bartlett, *Fundamentals of Fire Fighter Skills*, 1st Edition, 1st Printing, pages 971-973.
Answer: A

147. Reference: NFPA 1001, 5.3.14 and 5.3.14(A)
Delmar, *Firefighter's Handbook*, 2nd Edition, 1st Printing, page 646.
IFSTA, *Essentials of Fire Fighting*, 4th Edition, 1st Printing, page 595.
Jones and Bartlett, *Fundamentals of Fire Fighter Skills*, 1st Edition, 1st Printing, page 569.
Answer: B

148. Reference: NFPA 1001, 5.3.14 and 5.3.14(A)(B)
Delmar, *Firefighter's Handbook*, 2nd Edition, 1st Printing, pages 645-646.
IFSTA, *Essentials of Fire Fighting*, 4th Edition, 1st Printing, page 595.
Jones and Bartlett, *Fundamentals of Fire Fighter Skills*, 1st Edition, 1st Printing, page 569.
Answer: C

149. Reference: NFPA 1001, 5.3.14 and 5.3.14(A)(B)
Delmar, *Firefighter's Handbook*, 2nd Edition, 1st Printing, pages 645-646.
IFSTA, *Essentials of Fire Fighting*, 4th Edition, 1st Printing, page 617.
Jones and Bartlett, *Fundamentals of Fire Fighter Skills*, 1st Edition, 1st Printing, pages 571 and 573.
Answer: C

150. Reference: NFPA 1001, 5.3.14 and 5.3.14(A)(B)
Delmar, *Firefighter's Handbook*, 2nd Edition, 1st Printing, page 633.
IFSTA, *Essentials of Fire Fighting*, 4th Edition, 1st Printing, page 587.
Jones and Bartlett, *Fundamentals of Fire Fighter Skills*, 1st Edition, 1st Printing, page 554.
Answer: C

151. Reference: NFPA 1001, 5.3.14 and 5.3.14(A)(B)
Delmar, *Firefighter's Handbook*, 2nd Edition, 1st Printing, pages 633 and 649.
IFSTA, *Essentials of Fire Fighting*, 4th Edition, 1st Printing, page 587.
Jones and Bartlett, *Fundamentals of Fire Fighter Skills*, 1st Edition, 1st Printing, page 554.
Answer: C

152. Reference: NFPA 1001, 5.3.14 and 5.3.14(A)(B)
Delmar, *Firefighter's Handbook*, 2nd Edition, 1st Printing, pages 634, 637-639, and 642-646.
IFSTA, *Essentials of Fire Fighting*, 4th Edition, 1st Printing, pages 593 and 595.
Jones and Bartlett, *Fundamentals of Fire Fighter Skills*, 1st Edition, 1st Printing, pages 556, 562-563, and 569-573.
Answer: D

153. Reference: NFPA 1001, 5.3.14 and 5.3.14(A)(B)
Delmar, *Firefighter's Handbook*, 2nd Edition, 1st Printing, page 646.
IFSTA, *Essentials of Fire Fighting*, 4th Edition, 1st Printing, pages 615-616.
Jones and Bartlett, *Fundamentals of Fire Fighter Skills*, 1st Edition, 1st Printing, page 571.
Answer: B

154. Reference: NFPA 1001, 5.3.14 and 5.3.14(A)(B)
Delmar, *Firefighter's Handbook*, 2nd Edition, 1st Printing, page 633.
IFSTA, *Essentials of Fire Fighting*, 4th Edition, 1st Printing, page 587.
Jones and Bartlett, *Fundamentals of Fire Fighter Skills*, 1st Edition, 1st Printing, page 554.
Answer: C

155. Reference: NFPA 1001, 5.3.14 and 5.3.14(A)(B)
Delmar, *Firefighter's Handbook*, 2nd Edition, 1st Printing, page 639.
IFSTA, *Essentials of Fire Fighting*, 4th Edition, 1st Printing, page 587.
Jones and Bartlett, *Fundamentals of Fire Fighter Skills*, 1st Edition, 1st Printing, page 555.
Answer: B

156. Reference: NFPA 1001, 5.3.14 and 5.3.14(A)(B)
Delmar, *Firefighter's Handbook*, 2nd Edition, 1st Printing, pages 645-646.
IFSTA, *Essentials of Fire Fighting*, 4th Edition, 1st Printing, pages 592-593.
Jones and Bartlett, *Fundamentals of Fire Fighter Skills*, 1st Edition, 1st Printing, page 577.
Answer: D

157. Reference: NFPA 1001, 5.3.14 and 5.3.14(A)
Delmar, *Firefighter's Handbook*, 2nd Edition, 1st Printing, page 642.
Jones and Bartlett, *Fundamentals of Fire Fighter Skills*, 1st Edition, 1st Printing, page 555.
Answer: B

158. Reference: NFPA 1001, 5.3.14 and 5.3.14(A)
Delmar, *Firefighter's Handbook*, 2nd Edition, 1st Printing, page 331.
IFSTA, *Essentials of Fire Fighting*, 4th Edition, 1st Printing, page 578.
Jones and Bartlett, *Fundamentals of Fire Fighter Skills*, 1st Edition, 1st Printing, page 943.
Answer: C

159. Reference: NFPA 1001, 5.3.14 and 5.3.14(A)(B)
Delmar, *Firefighter's Handbook*, 2nd Edition, 1st Printing, page 326.
IFSTA, *Essentials of Fire Fighting*, 4th Edition, 1st Printing, page 584.
Jones and Bartlett, *Fundamentals of Fire Fighter Skills*, 1st Edition, 1st Printing, page 566.
Answer: B

160. Reference: NFPA 1001, 5.3.14 and 5.3.14(A)
Delmar, *Firefighter's Handbook*, 2nd Edition, 1st Printing, page 317.
IFSTA, *Essentials of Fire Fighting*, 4th Edition, 1st Printing, page 574.
Jones and Bartlett, *Fundamentals of Fire Fighter Skills*, 1st Edition, 1st Printing, page 938.
Answer: D

161. Reference: NFPA 1001, 5.3.14 and 5.3.14(A)
Delmar, *Firefighter's Handbook*, 2nd Edition, 1st Printing, page 317.
IFSTA, *Essentials of Fire Fighting*, 4th Edition, 1st Printing, page 574.
Jones and Bartlett, *Fundamentals of Fire Fighter Skills*, 1st Edition, 1st Printing, page 938.
Answer: A

162. Reference: NFPA 1001, 5.3.14 and 5.3.14(A)
Delmar, *Firefighter's Handbook*, 2nd Edition, 1st Printing, page 317.
IFSTA, *Essentials of Fire Fighting*, 4th Edition, 1st Printing, page 574.
Jones and Bartlett, *Fundamentals of Fire Fighter Skills*, 1st Edition, 1st Printing, page 938.
Answer: C

163. Reference: NFPA 1001, 5.3.14 and 5.3.14(A)(B)
Delmar, *Firefighter's Handbook*, 2nd Edition, 1st Printing, page 326.
IFSTA, *Essentials of Fire Fighting*, 4th Edition, 1st Printing, page 584.
Jones and Bartlett, *Fundamentals of Fire Fighter Skills*, 1st Edition, 1st Printing, page 566.
Answer: A

164. Reference: NFPA 1001, 5.3.15 and 5.3.15(A)
Delmar, *Firefighter's Handbook*, 2nd Edition, 1st Printing, page 223.
IFSTA, *Essentials of Fire Fighting*, 4th Edition, 1st Printing, page 415.
Jones and Bartlett, *Fundamentals of Fire Fighter Skills*, 1st Edition, 1st Printing, page 492.
Answer: C

165. Reference: NFPA 1001, 5.3.15 and 5.3.15(A)
Delmar, *Firefighter's Handbook*, 2nd Edition, 1st Printing, pages 210-211.
IFSTA, *Essentials of Fire Fighting*, 4th Edition, 1st Printing, page 379.
Jones and Bartlett, *Fundamentals of Fire Fighter Skills*, 1st Edition, 1st Printing, page 446.
Answer: B

166. Reference: NFPA 1001, 5.3.15 and 5.3.15(A)
Delmar, *Firefighter's Handbook*, 2nd Edition, 1st Printing, pages 291-292.
IFSTA, *Essentials of Fire Fighting*, 4th Edition, 1st Printing, pages 385-386.
Jones and Bartlett, *Fundamentals of Fire Fighter Skills*, 1st Edition, 1st Printing, pages 454-457.
Answer: D

167. Reference: NFPA 1001, 5.3.15 and 5.3.15(A)
Delmar, *Firefighter's Handbook*, 2nd Edition, 1st Printing, page 205.
IFSTA, *Essentials of Fire Fighting*, 4th Edition, 1st Printing, page 381.
Jones and Bartlett, *Fundamentals of Fire Fighter Skills*, 1st Edition, 1st Printing, page 448.
Answer: B

168. Reference: NFPA 1001, 5.3.15 and 5.3.15(A)
Delmar, *Firefighter's Handbook*, 2nd Edition, 1st Printing, pages 206-207.
IFSTA, *Essentials of Fire Fighting*, 4th Edition, 1st Printing, page 386.
Jones and Bartlett, *Fundamentals of Fire Fighter Skills*, 1st Edition, 1st Printing, pages 449-450.
Answer: A

169. Reference: NFPA 1001, 5.3.15 and 5.3.15(A)
Delmar, *Firefighter's Handbook*, 2nd Edition, 1st Printing, pages 206-208.
IFSTA, *Essentials of Fire Fighting*, 4th Edition, 1st Printing, pages 386-387.
Jones and Bartlett, *Fundamentals of Fire Fighter Skills*, 1st Edition, 1st Printing, pages 449-451.
Answer: D

170. Reference: NFPA 1001, 5.3.15 and 5.3.15(A)
Delmar, *Firefighter's Handbook*, 2nd Edition, 1st Printing, pages 214 and 951.
IFSTA, *Essentials of Fire Fighting*, 4th Edition, 1st Printing, page 386.
Jones and Bartlett, *Fundamentals of Fire Fighter Skills*, 1st Edition, 1st Printing, page 455.
Answer: B

171. Reference: NFPA 1001, 5.3.15 and 5.3.15(A)(B)
Delmar, *Firefighter's Handbook*, 2nd Edition, 1st Printing, page 211.
IFSTA, *Essentials of Fire Fighting*, 4th Edition, 1st Printing, page 391.
Jones and Bartlett, *Fundamentals of Fire Fighter Skills*, 1st Edition, 1st Printing, page 459.
Answer: B

172. Reference: NFPA 1001, 5.3.15 and 5.3.15(A)(B)
Delmar, *Firefighter's Handbook*, 2nd Edition, 1st Printing, page 212.
IFSTA, *Essentials of Fire Fighting*, 4th Edition, 1st Printing, page 391.
Jones and Bartlett, *Fundamentals of Fire Fighter Skills*, 1st Edition, 1st Printing, page 459.
Answer: B

173. Reference: NFPA 1001, 5.3.16 and 5.3.16(A)(B)
Delmar, *Firefighter's Handbook*, 2nd Edition, 1st Printing, page 194.
IFSTA, *Essentials of Fire Fighting*, 4th Edition, 1st Printing, page 134.
Jones and Bartlett, *Fundamentals of Fire Fighter Skills*, 1st Edition, 1st Printing, page 180.
Answer: B

174. Reference: NFPA 1001, 5.3.16 and 5.3.16(A)
Delmar, *Firefighter's Handbook*, 2nd Edition, 1st Printing, page 189.
IFSTA, *Essentials of Fire Fighting*, 4th Edition, 1st Printing, pages 133-135.
Jones and Bartlett, *Fundamentals of Fire Fighter Skills*, 1st Edition, 1st Printing, page 181.
Answer: A

175. Reference: NFPA 1001, 5.3.16 and 5.3.16(A)(B)
Delmar, *Firefighter's Handbook*, 2nd Edition, 1st Printing, page 195.
IFSTA, *Essentials of Fire Fighting*, 4th Edition, 1st Printing, pages 141-142.
Jones and Bartlett, *Fundamentals of Fire Fighter Skills*, 1st Edition, 1st Printing, page 202.
Answer: D

176. Reference: NFPA 1001, 5.3.16 and 5.3.16(A)
Delmar, *Firefighter's Handbook*, 2nd Edition, 1st Printing, page 189.
IFSTA, *Essentials of Fire Fighting*, 4th Edition, 1st Printing, pages 133 and 135.
Jones and Bartlett, *Fundamentals of Fire Fighter Skills*, 1st Edition, 1st Printing, page 181.
Answer: C

177. Reference: NFPA 1001, 5.3.16 and 5.3.16(A)(B)
Delmar, *Firefighter's Handbook*, 2nd Edition, 1st Printing, page 189.
IFSTA, *Essentials of Fire Fighting*, 4th Edition, 1st Printing, pages 133 and 135.
Jones and Bartlett, *Fundamentals of Fire Fighter Skills*, 1st Edition, 1st Printing, page 181.
Answer: D

178. Reference: NFPA 1001, 5.3.16 and 5.3.16(A)
Delmar, *Firefighter's Handbook*, 2nd Edition, 1st Printing, page 188.
IFSTA, *Essentials of Fire Fighting*, 4th Edition, 1st Printing, pages 126 and 130.
Jones and Bartlett, *Fundamentals of Fire Fighter Skills*, 1st Edition, 1st Printing, pages 185-186.
Answer: D

179. Reference: NFPA 1001, 5.3.16 and 5.3.16(A)(B)
Delmar, *Firefighter's Handbook*, 2nd Edition, 1st Printing, pages 185-186.
IFSTA, *Essentials of Fire Fighting*, 4th Edition, 1st Printing, page 58.
Jones and Bartlett, *Fundamentals of Fire Fighter Skills*, 1st Edition, 1st Printing, page 178.
Answer: B

180. Reference: NFPA 1001, 5.3.16 and 5.3.16(A)(B)
Delmar, *Firefighter's Handbook*, 2nd Edition, 1st Printing, page 186.
IFSTA, *Essentials of Fire Fighting*, 4th Edition, 1st Printing, page 59.
Jones and Bartlett, *Fundamentals of Fire Fighter Skills*, 1st Edition, 1st Printing, pages 188 and 193.
Answer: D

181. Reference: NFPA 1001, 5.3.16 and 5.3.16(A)
Delmar, *Firefighter's Handbook*, 2nd Edition, 1st Printing, page 190.
IFSTA, *Essentials of Fire Fighting*, 4th Edition, 1st Printing, page 139.
Jones and Bartlett, *Fundamentals of Fire Fighter Skills*, 1st Edition, 1st Printing, page 191.
Answer: C

182. Reference: NFPA 1001, 5.3.16 and 5.3.16(A)(B)
Delmar, *Firefighter's Handbook*, 2nd Edition, 1st Printing, page 187-188.
IFSTA, *Essentials of Fire Fighting*, 4th Edition, 1st Printing, page 128.
Jones and Bartlett, *Fundamentals of Fire Fighter Skills*, 1st Edition, 1st Printing, page 187.
Answer: B

183. Reference: NFPA 1001, 5.3.16 and 5.3.16(A)
Delmar, *Firefighter's Handbook*, 2nd Edition, 1st Printing, page 189.
IFSTA, *Essentials of Fire Fighting*, 4th Edition, 1st Printing, page 135.
Jones and Bartlett, *Fundamentals of Fire Fighter Skills*, 1st Edition, 1st Printing, page 181.
Answer: D

184. Reference: NFPA 1001, 5.3.16 and 5.3.16(A)(B)
Delmar, *Firefighter's Handbook*, 2nd Edition, 1st Printing, page 195.
IFSTA, *Essentials of Fire Fighting*, 4th Edition, 1st Printing, page 125.
Jones and Bartlett, *Fundamentals of Fire Fighter Skills*, 1st Edition, 1st Printing, page 177.
Answer: B

185. Reference: NFPA 1001, 5.3.16 and 5.3.16(A)(B)
Delmar, *Firefighter's Handbook*, 2nd Edition, 1st Printing, page 195.
Jones and Bartlett, *Fundamentals of Fire Fighter Skills*, 1st Edition, 1st Printing, page 195.
Answer: C

186. Reference: NFPA 1001, 5.3.16 and 5.3.16(A)
Delmar, *Firefighter's Handbook*, 2nd Edition, 1st Printing, page 187.
IFSTA, *Essentials of Fire Fighting*, 4th Edition, 1st Printing, pages 125 and 127.
Jones and Bartlett, *Fundamentals of Fire Fighter Skills*, 1st Edition, 1st Printing, page 185.
Answer: B

187. Reference: NFPA 1001, 5.3.16 and 5.3.16(A)
Delmar, *Firefighter's Handbook*, 2nd Edition, 1st Printing, page 188.
IFSTA, *Essentials of Fire Fighting*, 4th Edition, 1st Printing, page 127.
Jones and Bartlett, *Fundamentals of Fire Fighter Skills*, 1st Edition, 1st Printing, page 191.
Answer: A

188. Reference: NFPA 1001, 5.3.16 and 5.3.16(A)(B)
Delmar, *Firefighter's Handbook*, 2nd Edition, 1st Printing, page 188.
IFSTA, *Essentials of Fire Fighting*, 4th Edition, 1st Printing, page 131.
Jones and Bartlett, *Fundamentals of Fire Fighter Skills*, 1st Edition, 1st Printing, page 185.
Answer: C

189. Reference: NFPA 1001, 5.3.16 and 5.3.16(A)
Delmar, *Firefighter's Handbook*, 2nd Edition, 1st Printing, page 188.
IFSTA, *Essentials of Fire Fighting*, 4th Edition, 1st Printing, page 128.
Jones and Bartlett, *Fundamentals of Fire Fighter Skills*, 1st Edition, 1st Printing, page 188.
Answer: A

190. Reference: NFPA 1001, 5.3.18 and 5.3.18(A)(B)
Delmar, *Firefighter's Handbook*, 2nd Edition, 1st Printing, page 498.
IFSTA, *Essentials of Fire Fighting*, 4th Edition, 1st Printing, pages 210 and 540.
Jones and Bartlett, *Fundamentals of Fire Fighter Skills*, 1st Edition, 1st Printing, page 28.
Answer: A

191. Reference: NFPA 1001, 5.3.19 and 5.3.19(A)
Delmar, *Firefighter's Handbook*, 2nd Edition, 1st Printing, pages 595-599.
IFSTA, *Essentials of Fire Fighting*, 4th Edition, 1st Printing, page 552.
Jones and Bartlett, *Fundamentals of Fire Fighter Skills*, 1st Edition, 1st Printing, pages 609-611.
Answer: B

192. Reference: NFPA 1001, 5.3.19 and 5.3.19(A)
Delmar, *Firefighter's Handbook*, 2nd Edition, 1st Printing, page 621.
IFSTA, *Essentials of Fire Fighting*, 4th Edition, 1st Printing, page 556.
Jones and Bartlett, *Fundamentals of Fire Fighter Skills*, 1st Edition, 1st Printing, page 613.
Answer: B

193. Reference: NFPA 1001, 5.5.1 and 5.5.1(A)(B)
Delmar, *Firefighter's Handbook*, 2nd Edition, 1st Printing, pages 661-662.
IFSTA, *Essentials of Fire Fighting*, 4th Edition, 1st Printing, page 662.
Jones and Bartlett, *Fundamentals of Fire Fighter Skills*, 1st Edition, 1st Printing, page 651.
Answer: A

194. Reference: NFPA 1001, 5.5.1 and 5.5.1(A)(B)
Delmar, *Firefighter's Handbook*, 2nd Edition, 1st Printing, page 685.
IFSTA, *Essentials of Fire Fighting*, 4th Edition, 1st Printing, pages 662-663.
Jones and Bartlett, *Fundamentals of Fire Fighter Skills*, 1st Edition, 1st Printing, page 651.
Answer: D

195. Reference: NFPA 1001, 5.5.2 and 5.5.2(A)(B)
Delmar, *Firefighter's Handbook*, 2nd Edition, 1st Printing, page 680.
IFSTA, *Essentials of Fire Fighting*, 4th Edition, 1st Printing, page 671.
Jones and Bartlett, *Fundamentals of Fire Fighter Skills*, 1st Edition, 1st Printing, page 917.
Answer: D

196. Reference: NFPA 1001, 5.5.3 and 5.5.3(A)(B)
Delmar, *Firefighter's Handbook*, 2nd Edition, 1st Printing, page 177.
IFSTA, *Essentials of Fire Fighting*, 4th Edition, 1st Printing, page 113.
Jones and Bartlett, *Fundamentals of Fire Fighter Skills*, 1st Edition, 1st Printing, page 62.
Answer: B

197. Reference: NFPA 1001, 5.5.3 and 5.5.3(A)(B)
Delmar, *Firefighter's Handbook*, 2nd Edition, 1st Printing, page 154.
IFSTA, *Essentials of Fire Fighting*, 4th Edition, 1st Printing, page 106.
Jones and Bartlett, *Fundamentals of Fire Fighter Skills*, 1st Edition, 1st Printing, page 62.
Answer: B

198. Reference: NFPA 1001, 5.5.3 and 5.5.3(A)(B)
Delmar, *Firefighter's Handbook*, 2nd Edition, 1st Printing, pages 153-154.
IFSTA, *Essentials of Fire Fighting*, 4th Edition, 1st Printing, page 106.
Jones and Bartlett, *Fundamentals of Fire Fighter Skills*, 1st Edition, 1st Printing, page 62.
Answer: D

199. Reference: NFPA 1001, 5.5.3 and 5.5.3(A)(B)
Delmar, *Firefighter's Handbook*, 2nd Edition, 1st Printing, page 440.
IFSTA, *Essentials of Fire Fighting*, 4th Edition, 1st Printing, page 151.
Jones and Bartlett, *Fundamentals of Fire Fighter Skills*, 1st Edition, 1st Printing, page 252.
Answer: C

200. Reference: NFPA 1001, 5.5.4, 5.5.4(A)(B), 5.3.15 and 5.3.15(A)(B)
Delmar, *Firefighter's Handbook*, 2nd Edition, 1st Printing, page 243.
IFSTA, *Essentials of Fire Fighting*, 4th Edition, 1st Printing, pages 419-420.
Jones and Bartlett, *Fundamentals of Fire Fighter Skills*, 1st Edition, 1st Printing, page 501.
Answer: A

APPENDIX B

Examination II-1 Answer Key

Directions
Follow these steps carefully for completing the feedback step of Systematic Approach to Exam Preparation (SAEP):

1. After entering your scores, look up the answers for the examination items you missed, as well as those you guessed, even if you guessed correctly. If you are guessing, it means the answer is not perfectly clear. In this process we are committed to making you as knowledgeable as possible.
2. Enter the number of missed and guessed examination items in the blank on the personal progress plotter.
3. Highlight the answer in the reference materials. Once you have highlighted the answer, read the paragraph preceding and the paragraph following the one in which the correct answer is located. Now that you have highlighted the answer, enter the paragraph number and page number next to the guessed or missed examination item on your examination. Count any part of a paragraph at the beginning of the page as one paragraph until you reach the paragraph containing your highlighted answer. This step will help you locate and review your missed and guessed examination items later in the process. This step is **essential** to learning the material in context and by association. These learning techniques (context/association) are the very backbone of the SAEP approach.
4. Once you have completed the feedback step, you may proceed to the next examination.

1. Reference: NFPA 1001, 6.1.1.1 and 6.1.1.2
   Delmar, *Firefighter's Handbook*, 2nd Edition, 1st Printing, page 36.
   IFSTA, *Essentials of Fire Fighting*, 4th Edition, 1st Printing, pages 541 and 544.
   Jones and Bartlett, *Fundamentals of Fire Fighter Skills*, 1st Edition, 1st Printing, page 117.
   Answer: B

2. Reference: NFPA 1001, 6.1.1.1 and 6.1.1.2
   Delmar, *Firefighter's Handbook*, 2nd Edition, 1st Printing, page 36.
   IFSTA, *Essentials of Fire Fighting*, 4th Edition, 1st Printing, page 7.
   Jones and Bartlett, *Fundamentals of Fire Fighter Skills*, 1st Edition, 1st Printing, pages 8, 104.
   Answer: D

3. Reference: NFPA 1001, 6.1.1.1 and 6.1.1.2
   Delmar, *Firefighter's Handbook*, 2nd Edition, 1st Printing, page 23.
   IFSTA, *Essentials of Fire Fighting*, 4th Edition, 1st Printing, page 7.
   Jones and Bartlett, *Fundamentals of Fire Fighter Skills*, 1st Edition, 1st Printing, page 8.
   Answer: D

4. Reference: NFPA 1001, 6.1.1.1 and 6.1.1.2
   Delmar, *Firefighter's Handbook*, 2nd Edition, 1st Printing, page 34.
   IFSTA, *Essentials of Fire Fighting*, 4th Edition, 1st Printing, page 12.
   Jones and Bartlett, *Fundamentals of Fire Fighter Skills*, 1st Edition, 1st Printing, page 13.
   Answer: B

5. Reference: NFPA 1001, 6.1.1.1
   Delmar, *Firefighter's Handbook*, 2nd Edition, 1st Printing, page 35.
   IFSTA, *Essentials of Fire Fighting*, 4th Edition, 1st Printing, page 12.
   Jones and Bartlett, *Fundamentals of Fire Fighter Skills*, 1st Edition, 1st Printing, page 13.
   Answer: B

6. Reference: NFPA 1001, 6.1.1.1 and 6.1.1.2
   Delmar, *Firefighter's Handbook*, 2nd Edition, 1st Printing, page 39.
   IFSTA, *Essentials of Fire Fighting*, 4th Edition, 1st Printing, page 14.
   Jones and Bartlett, *Fundamentals of Fire Fighter Skills*, 1st Edition, 1st Printing, page 108.
   Answer: B

7. Reference: NFPA 1001, 6.1.1.1 and 6.1.1.2
   Delmar, *Firefighter's Handbook*, 2nd Edition, 1st Printing, page 39.
   IFSTA, *Essentials of Fire Fighting*, 4th Edition, 1st Printing, page 14.
   Jones and Bartlett, *Fundamentals of Fire Fighter Skills*, 1st Edition, 1st Printing, page 108.
   Answer: D

8. Reference: NFPA 1001, 6.1.1.1 and 6.1.1.2
   Delmar, *Firefighter's Handbook*, 1st Edition, 1st Printing, page 39.
   IFSTA, *Essentials of Fire Fighting*, 4th Edition, 1st Printing, pages 14-15.
   Jones and Bartlett, *Fundamentals of Fire Fighter Skills*, 1st Edition, 1st Printing, page 108.
   Answer: A

9. Reference: NFPA 1001, 6.1.1.1
   Delmar, *Firefighter's Handbook*, 2nd Edition, 1st Printing, page 109.
   IFSTA, *Essentials of Fire Fighting*, 4th Edition, 1st Printing, page 21.
   Jones and Bartlett, *Fundamentals of Fire Fighter Skills*, 1st Edition, 1st Printing, page 24.
   Answer: B

10. Reference: NFPA 1001, 6.1.1.1 and 6.1.1.2
Delmar, *Firefighter's Handbook*, 2nd Edition, 1st Printing, page 39.
IFSTA, *Essentials of Fire Fighting*, 4th Edition, 1st Printing, page 649.
Jones and Bartlett, *Fundamentals of Fire Fighter Skills*, 1st Edition, 1st Printing, page 106.
Answer: D

11. Reference: NFPA 1001, 6.1.1.1 and 6.1.1.2
Delmar, *Firefighter's Handbook*, 2nd Edition, 1st Printing, page 41.
IFSTA, *Essentials of Fire Fighting*, 4th Edition, 1st Printing, page 15.
Jones and Bartlett, *Fundamentals of Fire Fighter Skills*, 1st Edition, 1st Printing, page 111.
Answer: C

12. Reference: NFPA 1001, 6.1.1.1 and 6.1.1.2
Delmar, *Firefighter's Handbook*, 2nd Edition, 1st Printing, page 38.
IFSTA, *Essentials of Fire Fighting*, 4th Edition, 1st Printing, page 14.
Jones and Bartlett, *Fundamentals of Fire Fighter Skills*, 1st Edition, 1st Printing, page 104.
Answer: D

13. Reference: NFPA 1001, 6.1.1.1 and 6.1.1.2
Delmar, *Firefighter's Handbook*, 2nd Edition, 1st Printing, page 68.
IFSTA, *Essentials of Fire Fighting*, 4th Edition, 1st Printing, page 648.
Jones and Bartlett, *Fundamentals of Fire Fighter Skills*, 1st Edition, 1st Printing, page 87.
Answer: D

14. Reference: NFPA 1001, 6.1.1.1
Delmar, *Firefighter's Handbook*, 2nd Edition, 1st Printing, page 314.
Jones and Bartlett, *Fundamentals of Fire Fighter Skills*, 1st Edition, 1st Printing, page 3.
Answer: D

15. Reference: NFPA 1001, 6.1.1.1, 6.1.1.2, and 6.3.2(A)
Delmar, *Firefighter's Handbook*, 2nd Edition, 1st Printing, page 118.
IFSTA, *Essentials of Fire Fighting*, 4th Edition, 1st Printing, pages 28-29.
Jones and Bartlett, *Fundamentals of Fire Fighter Skills*, 1st Edition, 1st Printing, page 383.
Answer: C

16. Reference: NFPA 1001, 6.2.1 and 6.2.1(A)(B)
Delmar, *Firefighter's Handbook*, 2nd Edition, 1st Printing, page 69.
IFSTA, *Essentials of Fire Fighting*, 4th Edition, 1st Printing, page 651.
Jones and Bartlett, *Fundamentals of Fire Fighter Skills*, 1st Edition, 1st Printing, page 89.
Answer: D

17. Reference: NFPA 1001, 6.2.1 and 6.2.1(A)(B)
Delmar, *Firefighter's Handbook*, 2nd Edition, 1st Printing, page 69.
IFSTA, *Essentials of Fire Fighting*, 4th Edition, 1st Printing, page 651.
Jones and Bartlett, *Fundamentals of Fire Fighter Skills*, 1st Edition, 1st Printing, page 89.
Answer: A

18. Reference: NFPA 1001, 6.2.1 and 6.2.1(A)
Delmar, *Firefighter's Handbook*, 2nd Edition, 1st Printing, page 69.
IFSTA, *Essentials of Fire Fighting*, 4th Edition, 1st Printing, page 651.
Jones and Bartlett, *Fundamentals of Fire Fighter Skills*, 1st Edition, 1st Printing, page 89.
Answer: D

19. Reference: NFPA 1001, 6.2.2, 6.2.2(A)(B), 6.3.2 and 6.3.2(A)(B)
Delmar, *Firefighter's Handbook*, 2nd Edition, 1st Printing, page 67.
IFSTA, *Essentials of Fire Fighting*, 4th Edition, 1st Printing, page 649.
Jones and Bartlett, *Fundamentals of Fire Fighter Skills*, 1st Edition, 1st Printing, page 89.
Answer: D

20. Reference: NFPA 1001, 6.2.2 and 6.2.2(A)(B)
Delmar, *Firefighter's Handbook*, 2nd Edition, 1st Printing, page 67.
IFSTA, *Essentials of Fire Fighting*, 4th Edition, 1st Printing, page 651.
Jones and Bartlett, *Fundamentals of Fire Fighter Skills*, 1st Edition, 1st Printing, page 89.
Answer: A

21. Reference: NFPA 1001, 6.2.2 and 6.2.2(A)(B)
Delmar, *Firefighter's Handbook*, 2nd Edition, 1st Printing, page 68.
IFSTA, *Essentials of Fire Fighting*, 4th Edition, 1st Printing, page 648.
Jones and Bartlett, *Fundamentals of Fire Fighter Skills*, 1st Edition, 1st Printing, page 87.
Answer: D

22. Reference: NFPA 1001, 6.2.2 and 6.2.2(A)(B)
Delmar, *Firefighter's Handbook*, 2nd Edition, 1st Printing, page 63.
IFSTA, *Essentials of Fire Fighting*, 4th Edition, 1st Printing, page 645.
Jones and Bartlett, *Fundamentals of Fire Fighter Skills*, 1st Edition, 1st Printing, page 83.
Answer: C

23. Reference: NFPA 1001, 6.2.2 and 6.2.2(A)(B)
IFSTA, *Essentials of Fire Fighting*, 4th Edition, 1st Printing, page 636.
Answer: C

24. Reference: NFPA 1001, 6.2.2 and 6.2.2(A)(B)
Delmar, *Firefighter's Handbook*, 2nd Edition, 1st Printing, page 67.
IFSTA, *Essentials of Fire Fighting*, 4th Edition, 1st Printing, page 648.
Jones and Bartlett, *Fundamentals of Fire Fighter Skills*, 1st Edition, 1st Printing, page 85.
Answer: D

APPENDIX B, EXAMINATION II-1 ANSWER KEY

25. Reference: NFPA 1001, 6.2.2 and 6.2.2(A)(B)
Delmar, *Firefighter's Handbook*, 2nd Edition, 1st Printing, page 54.
IFSTA, *Essentials of Fire Fighting*, 4th Edition, 1st Printing, page 641.
Jones and Bartlett, *Fundamentals of Fire Fighter Skills*, 1st Edition, 1st Printing, pages 77-78.
Answer: A

26. Reference: NFPA 1001, 6.2.2 and 6.2.2(A)
Delmar, *Firefighter's Handbook*, 2nd Edition, 1st Printing, page 49.
IFSTA, *Essentials of Fire Fighting*, 4th Edition, 1st Printing, page 636.
Jones and Bartlett, *Fundamentals of Fire Fighter Skills*, 1st Edition, 1st Printing, page 72.
Answer: B

27. Reference: NFPA 1001, 6.2.2 and 6.2.2(A)(B)
Delmar, *Firefighter's Handbook*, 2nd Edition, 1st Printing, page 68.
IFSTA, *Essentials of Fire Fighting*, 4th Edition, 1st Printing, page 646.
Jones and Bartlett, *Fundamentals of Fire Fighter Skills*, 1st Edition, 1st Printing, page 87.
Answer: B

28. Reference: NFPA 1001, 6.3.1 and 6.3.1(A)(B)
IFSTA, *Essentials of Fire Fighting*, 4th Edition, 1st Printing, page 516.
Jones and Bartlett, *Fundamentals of Fire Fighter Skills*, 1st Edition, 1st Printing, page 526.
Answer: C

29. Reference: NFPA 1001, 6.3.1 and 6.3.1(A)(B)
Delmar, *Firefighter's Handbook*, 2nd Edition, 1st Printing, page 298.
IFSTA, *Essentials of Fire Fighting*, 4th Edition, 1st Printing, page 510.
Jones and Bartlett, *Fundamentals of Fire Fighter Skills*, 1st Edition, 1st Printing, page 527.
Answer: B

30. Reference: NFPA 1001, 6.3.1 and 6.3.1(A)(B)
Delmar, *Firefighter's Handbook*, 2nd Edition, 1st Printing, page 295.
IFSTA, *Essentials of Fire Fighting*, 4th Edition, 1st Printing, page 499.
Jones and Bartlett, *Fundamentals of Fire Fighter Skills*, 1st Edition, 1st Printing, page 525.
Answer: A

31. Reference: NFPA 1001, 6.3.1 and 6.3.1(A)(B)
Delmar, *Firefighter's Handbook*, 2nd Edition, 1st Printing, page 296.
IFSTA, *Essentials of Fire Fighting*, 4th Edition, 1st Printing, page 500.
Jones and Bartlett, *Fundamentals of Fire Fighter Skills*, 1st Edition, 1st Printing, page 525.
Answer: C

32. Reference: NFPA 1001, 6.3.1 and 6.3.1(A)(B)
Delmar, *Firefighter's Handbook*, 2nd Edition, 1st Printing, page 296.
IFSTA, *Essentials of Fire Fighting*, 4th Edition, 1st Printing, page 503.
Jones and Bartlett, *Fundamentals of Fire Fighter Skills*, 1st Edition, 1st Printing, page 526.
Answer: B

33. Reference: NFPA 1001, 6.3.1 and 6.3.1(A)(B)
Delmar, *Firefighter's Handbook*, 2nd Edition, 1st Printing, page 295.
IFSTA, *Essentials of Fire Fighting*, 4th Edition, 1st Printing, page 498.
Jones and Bartlett, *Fundamentals of Fire Fighter Skills*, 1st Edition, 1st Printing, page 639.
Answer: C

34. Reference: NFPA 1001, 6.3.1 and 6.3.1(A)(B)
Delmar, *Firefighter's Handbook*, 2nd Edition, 1st Printing, pages 297 and 601.
IFSTA, *Essentials of Fire Fighting*, 4th Edition, 1st Printing, page 499.
Jones and Bartlett, *Fundamentals of Fire Fighter Skills*, 1st Edition, 1st Printing, page 187.
Answer: C

35. Reference: NFPA 1001, 6.3.1 and 6.3.1(A)(B)
Delmar, *Firefighter's Handbook*, 2nd Edition, 1st Printing, pages 297 and 601.
IFSTA, *Essentials of Fire Fighting*, 4th Edition, 1st Printing, pages 498-499.
Jones and Bartlett, *Fundamentals of Fire Fighter Skills*, 1st Edition, 1st Printing, page 526.
Answer: B

36. Reference: NFPA 1001, 6.3.1 and 6.3.1(A)(B)
Delmar, *Firefighter's Handbook*, 2nd Edition, 1st Printing, page 297.
IFSTA, *Essentials of Fire Fighting*, 4th Edition, 1st Printing, page 507.
Jones and Bartlett, *Fundamentals of Fire Fighter Skills*, 1st Edition, 1st Printing, page 527.
Answer: B

37. Reference: NFPA 1001, 6.3.2 and 6.3.2(A)(B)
Delmar, *Firefighter's Handbook*, 2nd Edition, 1st Printing, page 612.
IFSTA, *Essentials of Fire Fighting*, 4th Edition, 1st Printing, page 538.
Jones and Bartlett, *Fundamentals of Fire Fighter Skills*, 1st Edition, 1st Printing, page 642.
Answer: D

38. Reference: NFPA 1001, 6.3.2 and 6.3.2(A)(B)
Delmar, *Firefighter's Handbook*, 2nd Edition, 1st Printing, page 362.
IFSTA, *Essentials of Fire Fighting*, 4th Edition, 1st Printing, page 359.
Jones and Bartlett, *Fundamentals of Fire Fighter Skills*, 1st Edition, 1st Printing, page 161.
Answer: C

39. Reference: NFPA 1001, 6.3.2 and 6.3.2(A)(B)
Delmar, *Firefighter's Handbook*, 2nd Edition, 1st Printing, page 353.
IFSTA, *Essentials of Fire Fighting*, 4th Edition, 1st Printing, page 65.
Jones and Bartlett, *Fundamentals of Fire Fighter Skills*, 1st Edition, 1st Printing, page 411.
Answer: B

40. Reference: NFPA 1001, 6.3.2 and 6.3.2(A)(B)
IFSTA, *Essentials of Fire Fighting*, 4th Edition, 1st Printing, page 364.
Jones and Bartlett, *Fundamentals of Fire Fighter Skills*, 1st Edition, 1st Printing, pages 130-131.
Answer: C

41. Reference: NFPA 1001, 6.3.2 and 6.3.2(A)(B)
Delmar, *Firefighter's Handbook*, 2nd Edition, 1st Printing, page 353.
IFSTA, *Essentials of Fire Fighting*, 4th Edition, 1st Printing, page 65.
Jones and Bartlett, *Fundamentals of Fire Fighter Skills*, 1st Edition, 1st Printing, page 411.
Answer: C

42. Reference: NFPA 1001, 6.3.2 and 6.3.2(A)(B)
Delmar, *Firefighter's Handbook*, 2nd Edition, 1st Printing, page 350.
IFSTA, *Essentials of Fire Fighting*, 4th Edition, 1st Printing, page 70.
Jones and Bartlett, *Fundamentals of Fire Fighter Skills*, 1st Edition, 1st Printing, page 150.
Answer: D

43. Reference: NFPA 1001, 6.3.2 and 6.3.2(A)(B)
Delmar, *Firefighter's Handbook*, 2nd Edition, 1st Printing, page 353.
IFSTA, *Essentials of Fire Fighting*, 4th Edition, 1st Printing, page 65.
Jones and Bartlett, *Fundamentals of Fire Fighter Skills*, 1st Edition, 1st Printing, page 154.
Answer: A

44. Reference: NFPA 1001, 6.3.2 and 6.3.2(A)(B)
Delmar, *Firefighter's Handbook*, 2nd Edition, 1st Printing, page 357.
IFSTA, *Essentials of Fire Fighting*, 4th Edition, 1st Printing, page 67.
Jones and Bartlett, *Fundamentals of Fire Fighter Skills*, 1st Edition, 1st Printing, page 158.
Answer: D

45. Reference: NFPA 1001, 6.3.2 and 6.3.2(A)(B)
Delmar, *Firefighter's Handbook*, 2nd Edition, 1st Printing, page 355.
IFSTA, *Essentials of Fire Fighting*, 4th Edition, 1st Printing, page 67.
Jones and Bartlett, *Fundamentals of Fire Fighter Skills*, 1st Edition, 1st Printing, page 157.
Answer: C

46. Reference: NFPA 1001, 6.3.2 and 6.3.2(A)(B)
Delmar, *Firefighter's Handbook*, 2nd Edition, 1st Printing, page 361.
IFSTA, *Essentials of Fire Fighting*, 4th Edition, 1st Printing, pages 75-76.
Jones and Bartlett, *Fundamentals of Fire Fighter Skills*, 1st Edition, 1st Printing, page 427.
Answer: D

47. Reference: NFPA 1001, 6.3.2 and 6.3.2(A)(B)
Delmar, *Firefighter's Handbook*, 2nd Edition, 1st Printing, page 361.
IFSTA, *Essentials of Fire Fighting*, 4th Edition, 1st Printing, page 74.
Jones and Bartlett, *Fundamentals of Fire Fighter Skills*, 1st Edition, 1st Printing, page 427.
Answer: B

48. Reference: NFPA 1001, 6.3.2 and 6.3.2(A)(B)
Delmar, *Firefighter's Handbook*, 2nd Edition, 1st Printing, page 361.
IFSTA, *Essentials of Fire Fighting*, 4th Edition, 1st Printing, page 76.
Jones and Bartlett, *Fundamentals of Fire Fighter Skills*, 1st Edition, 1st Printing, page 427.
Answer: C

49. Reference: NFPA 1001, 6.3.2 and 6.3.2(A)(B)
Delmar, *Firefighter's Handbook*, 2nd Edition, 1st Printing, page 364.
IFSTA, *Essentials of Fire Fighting*, 4th Edition, 1st Printing, pages 73-74.
Jones and Bartlett, *Fundamentals of Fire Fighter Skills*, 1st Edition, 1st Printing, page 2.
Answer: D

50. Reference: NFPA 1001, 6.3.2 and 6.3.2(A)(B)
Delmar, *Firefighter's Handbook*, 2nd Edition, 1st Printing, page 543.
IFSTA, *Essentials of Fire Fighting*, 4th Edition, 1st Printing, page 71.
Jones and Bartlett, *Fundamentals of Fire Fighter Skills*, 1st Edition, 1st Printing, page 151.
Answer: A

51. Reference: NFPA 1001, 6.3.2 and 6.3.2(A)(B)
IFSTA, *Essentials of Fire Fighting*, 4th Edition, 1st Printing, page 68.
Jones and Bartlett, *Fundamentals of Fire Fighter Skills*, 1st Edition, 1st Printing, page 165.
Answer: B

52. Reference: NFPA 1001, 6.3.2 and 6.3.2(A)(B)
Delmar, *Firefighter's Handbook*, 2nd Edition, 1st Printing, page 520.
IFSTA, *Essentials of Fire Fighting*, 4th Edition, 1st Printing, pages 240-241.
Jones and Bartlett, *Fundamentals of Fire Fighter Skills*, 1st Edition, 1st Printing, page 299.
Answer: D

53. Reference: NFPA 1001, 6.3.2 and 6.3.2(A)(B)
Delmar, *Firefighter's Handbook*, 2nd Edition, 1st Printing, page 83.
IFSTA, *Essentials of Fire Fighting*, 4th Edition, 1st Printing, page 47.
Jones and Bartlett, *Fundamentals of Fire Fighter Skills*, 1st Edition, 1st Printing, page 127.
Answer: C

54. Reference: NFPA 1001, 6.3.2 and 6.3.2(A)(B)
Delmar, *Firefighter's Handbook*, 2nd Edition, 1st Printing, page 800.
IFSTA, *Essentials of Fire Fighting*, 4th Edition, 1st Printing, page 48.
Jones and Bartlett, *Fundamentals of Fire Fighter Skills*, 1st Edition, 1st Printing, page 133.
Answer: C

55. Reference: NFPA 1001, 6.3.3 and 6.3.3(A)(B)
Delmar, *Firefighter's Handbook*, 2nd Edition, 1st Printing, page 866.
IFSTA, *Essentials of Fire Fighting*, 4th Edition, 1st Printing, page 530.
Jones and Bartlett, *Fundamentals of Fire Fighter Skills*, 1st Edition, 1st Printing, page 641.
Answer: B

56. Reference: NFPA 1001, 6.3.3 and 6.3.3(A)(B)
Delmar, *Firefighter's Handbook*, 2nd Edition, 1st Printing, page 866.
IFSTA, *Essentials of Fire Fighting*, 4th Edition, 1st Printing, page 531.
Jones and Bartlett, *Fundamentals of Fire Fighter Skills*, 1st Edition, 1st Printing, page 641.
Answer: B

57. Reference: NFPA 1001, 6.3.3 and 6.3.3(A)(B)
Delmar, *Firefighter's Handbook*, 2nd Edition, 1st Printing, page 866.
IFSTA, *Essentials of Fire Fighting*, 4th Edition, 1st Printing, page 534.
Jones and Bartlett, *Fundamentals of Fire Fighter Skills*, 1st Edition, 1st Printing, page 641.
Answer: D

58. Reference: NFPA 1001, 6.3.3 and 6.3.3(A)(B)
Delmar, *Firefighter's Handbook*, 2nd Edition, 1st Printing, page 866.
IFSTA, *Essentials of Fire Fighting*, 4th Edition, 1st Printing, page 530.
Jones and Bartlett, *Fundamentals of Fire Fighter Skills*, 1st Edition, 1st Printing, page 641.
Answer: A

59. Reference: NFPA 1001, 6.3.4 and 6.3.4(A)(B)
Delmar, *Firefighter's Handbook*, 2nd Edition, 1st Printing, page 652.
IFSTA, *Essentials of Fire Fighting*, 4th Edition, 1st Printing, pages 623-625.
Jones and Bartlett, *Fundamentals of Fire Fighter Skills*, 1st Edition, 1st Printing, page 971-973.
Answer: D

60. Reference: NFPA 1001, 6.3.4 and 6.3.4(A)(B)
Delmar, *Firefighter's Handbook*, 2nd Edition, 1st Printing, page 652.
IFSTA, *Essentials of Fire Fighting*, 4th Edition, 1st Printing, page 622.
Jones and Bartlett, *Fundamentals of Fire Fighter Skills*, 1st Edition, 1st Printing, page 972.
Answer: B

61. Reference: NFPA 1001, 6.3.4 and 6.3.4(A)(B)
Delmar, *Firefighter's Handbook*, 1st Edition, 1st Printing, page 653.
IFSTA, *Essentials of Fire Fighting*, 4th Edition, 1st Printing, page 629.
Jones and Bartlett, *Fundamentals of Fire Fighter Skills*, 1st Edition, 1st Printing, page 967.
Answer: B

62. Reference: NFPA 1001, 6.3.4 and 6.3.4(A)(B)
Delmar, *Firefighter's Handbook*, 2nd Edition, 1st Printing, page 653.
IFSTA, *Essentials of Fire Fighting*, 4th Edition, 1st Printing, page 629.
Jones and Bartlett, *Fundamentals of Fire Fighter Skills*, 1st Edition, 1st Printing, page 967.
Answer: D

63. Reference: NFPA 1001, 6.3.4 and 6.3.4(A)(B)
Delmar, *Firefighter's Handbook*, 2nd Edition, 1st Printing, page 650.
IFSTA, *Essentials of Fire Fighting*, 4th Edition, 1st Printing, page 597.
Jones and Bartlett, *Fundamentals of Fire Fighter Skills*, 1st Edition, 1st Printing, page 576.
Answer: D

64. Reference: NFPA 1001, 6.3.4 and 6.3.4(A)(B)
Delmar, *Firefighter's Handbook*, 2nd Edition, 1st Printing, page 653.
IFSTA, *Essentials of Fire Fighting*, 4th Edition, 1st Printing, pages 621-622.
Jones and Bartlett, *Fundamentals of Fire Fighter Skills*, 1st Edition, 1st Printing, page 967.
Answer: D

65. Reference: NFPA 1001, 6.3.4 and 6.3.4(A)(B)
Delmar, *Firefighter's Handbook*, 1st Edition, 1st Printing, page 649.
IFSTA, *Essentials of Fire Fighting*, 4th Edition, 1st Printing, page 627.
Jones and Bartlett, *Fundamentals of Fire Fighter Skills*, 1st Edition, 1st Printing, page 574.
Answer: D

66. Reference: NFPA 1001, 6.3.4 and 6.3.4(A)(B)
Delmar, *Firefighter's Handbook*, 2nd Edition, 1st Printing, pages 652-653.
IFSTA, *Essentials of Fire Fighting*, 4th Edition, 1st Printing, page 621.
Jones and Bartlett, *Fundamentals of Fire Fighter Skills*, 1st Edition, 1st Printing, page 973.
Answer: A

67. Reference: NFPA 1001, 6.4.1 and 6.4.1(A)(B)
Delmar, *Firefighter's Handbook*, 2nd Edition, 1st Printing, page 487.
IFSTA, *Essentials of Fire Fighting*, 4th Edition, 1st Printing, pages 198-199.
Jones and Bartlett, *Fundamentals of Fire Fighter Skills*, 1st Edition, 1st Printing, page 738.
Answer: A

68. Reference: NFPA 1001, 6.4.1 and 6.4.1(A)(B)
Delmar, *Firefighter's Handbook*, 2nd Edition, 1st Printing, pages 487-488.
IFSTA, *Essentials of Fire Fighting*, 4th Edition, 1st Printing, page 192.
Jones and Bartlett, *Fundamentals of Fire Fighter Skills*, 1st Edition, 1st Printing, page 738.
Answer: D

69. Reference: NFPA 1001, 6.4.1 and 6.4.1(A)(B)
Delmar, *Firefighter's Handbook*, 2nd Edition, 1st Printing, page 488.
IFSTA, *Essentials of Fire Fighting*, 4th Edition, 1st Printing, page 200.
Jones and Bartlett, *Fundamentals of Fire Fighter Skills*, 1st Edition, 1st Printing, page 740.
Answer: B

70. Reference: NFPA 1001, 6.4.1 and 6.4.1(A)(B)
Delmar, *Firefighter's Handbook*, 2nd Edition, 1st Printing, page 486.
IFSTA, *Essentials of Fire Fighting*, 4th Edition, 1st Printing, page 197.
Jones and Bartlett, *Fundamentals of Fire Fighter Skills*, 1st Edition, 1st Printing, page 737.
Answer: D

71. Reference: NFPA 1001, 6.4.2 and 6.4.2(A)(B)
Delmar, *Firefighter's Handbook*, 2nd Edition, 1st Printing, page 498.
IFSTA, *Essentials of Fire Fighting*, 4th Edition, 1st Printing, page 208.
Jones and Bartlett, *Fundamentals of Fire Fighter Skills*, 1st Edition, 1st Printing, page 765.
Answer: A

72. Reference: NFPA 1001, 6.4.2 and 6.4.2(A)(B)
Delmar, *Firefighter's Handbook*, 2nd Edition, 1st Printing, page 482.
IFSTA, *Essentials of Fire Fighting*, 4th Edition, 1st Printing, page 188.
Jones and Bartlett, *Fundamentals of Fire Fighter Skills*, 1st Edition, 1st Printing, page 301.
Answer: B

73. Reference: NFPA 1001, 6.4.2 and 6.4.2(A)(B)
Delmar, *Firefighter's Handbook*, 2nd Edition, 1st Printing, page 484.
IFSTA, *Essentials of Fire Fighting*, 4th Edition, 1st Printing, page 196.
Jones and Bartlett, *Fundamentals of Fire Fighter Skills*, 1st Edition, 1st Printing, page 739.
Answer: A

74. Reference: NFPA 1001, 6.4.2 and 6.4.2(A)
Delmar, *Firefighter's Handbook*, 2nd Edition, 1st Printing, page 498.
IFSTA, *Essentials of Fire Fighting*, 4th Edition, 1st Printing, page 549.
Jones and Bartlett, *Fundamentals of Fire Fighter Skills*, 1st Edition, 1st Printing, page 762.
Answer: C

75. Reference: NFPA 1001, 6.4.2 and 6.4.2(A)(B)
Delmar, *Firefighter's Handbook*, 2nd Edition, 1st Printing, page 496.
IFSTA, *Essentials of Fire Fighting*, 4th Edition, 1st Printing, page 208.
Jones and Bartlett, *Fundamentals of Fire Fighter Skills*, 1st Edition, 1st Printing, page 765.
Answer: A

76. Reference: NFPA 1001, 6.4.2 and 6.4.2(A)(B)
Delmar, *Firefighter's Handbook*, 2nd Edition, 1st Printing, page 497.
IFSTA, *Essentials of Fire Fighting*, 4th Edition, 1st Printing, pages 87, 92, and 549.
Jones and Bartlett, *Fundamentals of Fire Fighter Skills*, 1st Edition, 1st Printing, page 762.
Answer: A

77. Reference: NFPA 1001, 6.5.1 and 6.5.1(A)(B)
Delmar, *Firefighter's Handbook*, 2nd Edition, 1st Printing, page 57.
IFSTA, *Essentials of Fire Fighting*, 4th Edition, 1st Printing, page 570.
Jones and Bartlett, *Fundamentals of Fire Fighter Skills*, 1st Edition, 1st Printing, page 927.
Answer: D

78. Reference: NFPA 1001, 6.5.1 and 6.5.1(A)(B)
Delmar, *Firefighter's Handbook*, 2nd Edition, 1st Printing, page 58.
IFSTA, *Essentials of Fire Fighting*, 4th Edition, 1st Printing, pages 637-638.
Jones and Bartlett, *Fundamentals of Fire Fighter Skills*, 1st Edition, 1st Printing, page 78.
Answer: B

79. Reference: NFPA 1001, 6.5.1 and 6.5.1(A)(B)
Delmar, *Firefighter's Handbook*, 2nd Edition, 1st Printing, page 205.
IFSTA, *Essentials of Fire Fighting*, 4th Edition, 1st Printing, pages 380-381.
Jones and Bartlett, *Fundamentals of Fire Fighter Skills*, 1st Edition, 1st Printing, page 448.
Answer: B

80. Reference: NFPA 1001, 6.5.1 and 6.5.1(A)(B)
Delmar, *Firefighter's Handbook*, 2nd Edition, 1st Printing, page 309.
IFSTA, *Essentials of Fire Fighting*, 4th Edition, 1st Printing, page 570.
Jones and Bartlett, *Fundamentals of Fire Fighter Skills*, 1st Edition, 1st Printing, page 927.
Answer: D

81. Reference: NFPA 1001, 6.5.1 and 6.5.1(A)(B)
Delmar, *Firefighter's Handbook*, 2nd Edition, 1st Printing, page 310.
IFSTA, *Essentials of Fire Fighting*, 4th Edition, 1st Printing, page 560.
Jones and Bartlett, *Fundamentals of Fire Fighter Skills*, 1st Edition, 1st Printing, page 928.
Answer: B

82. Reference: NFPA 1001, 6.5.1 and 6.5.1(A)(B)
Delmar, *Firefighter's Handbook*, 2nd Edition, 1st Printing, page 311.
IFSTA, *Essentials of Fire Fighting*, 4th Edition, 1st Printing, page 563.
Jones and Bartlett, *Fundamentals of Fire Fighter Skills*, 1st Edition, 1st Printing, page 928.
Answer: C

83. Reference: NFPA 1001, 6.5.1 and 6.5.1(A)
Delmar, *Firefighter's Handbook*, 2nd Edition, 1st Printing, page 323.
IFSTA, *Essentials of Fire Fighting*, 4th Edition, 1st Printing, page 579.
Jones and Bartlett, *Fundamentals of Fire Fighter Skills*, 1st Edition, 1st Printing, page 942.
Answer: A

84. Reference: NFPA 1001, 6.5.1 and 6.5.1(A)
Delmar, *Firefighter's Handbook*, 2nd Edition, 1st Printing, page 319.
IFSTA, *Essentials of Fire Fighting*, 4th Edition, 1st Printing, page 580.
Jones and Bartlett, *Fundamentals of Fire Fighter Skills*, 1st Edition, 1st Printing, page 944.
Answer: B

85. Reference: NFPA 1001, 6.5.1 and 6.5.1(A)
Delmar, *Firefighter's Handbook*, 2nd Edition, 1st Printing, pages 318-321.
IFSTA, *Essentials of Fire Fighting*, 4th Edition, 1st Printing, page 580.
Jones and Bartlett, *Fundamentals of Fire Fighter Skills*, 1st Edition, 1st Printing, page 943.
Answer: D

86. Reference: NFPA 1001, 6.5.1 and 6.5.1(A)(B)
Delmar, *Firefighter's Handbook*, 2nd Edition, 1st Printing, page 324.
IFSTA, *Essentials of Fire Fighting*, 4th Edition, 1st Printing, pages 384 and 577.
Jones and Bartlett, *Fundamentals of Fire Fighter Skills*, 1st Edition, 1st Printing, page 940.
Answer: B

87. Reference: NFPA 1001, 6.5.1 and 6.5.1(A)(B)
Delmar, *Firefighter's Handbook*, 2nd Edition, 1st Printing, page 320.
IFSTA, *Essentials of Fire Fighting*, 4th Edition, 1st Printing, pages 581-582.
Jones and Bartlett, *Fundamentals of Fire Fighter Skills*, 1st Edition, 1st Printing, page 946.
Answer: C

88. Reference: NFPA 1001, 6.5.1 and 6.5.1(A)(B)
Delmar, *Firefighter's Handbook*, 2nd Edition, 1st Printing, page 322.
IFSTA, *Essentials of Fire Fighting*, 4th Edition, 1st Printing, pages 577-579.
Jones and Bartlett, *Fundamentals of Fire Fighter Skills*, 1st Edition, 1st Printing, page 942.
Answer: C

89. Reference: NFPA 1001, 6.5.1 and 6.5.1(A)(B)
Delmar, *Firefighter's Handbook*, 2nd Edition, 1st Printing, page 318.
IFSTA, *Essentials of Fire Fighting*, 4th Edition, 1st Printing, pages 577-579.
Jones and Bartlett, *Fundamentals of Fire Fighter Skills*, 1st Edition, 1st Printing, page 940.
Answer: D

90. Reference: NFPA 1001, 6.5.1 and 6.5.1(A)(B)
Delmar, *Firefighter's Handbook*, 2nd Edition, 1st Printing, pages 318 and 323.
IFSTA, *Essentials of Fire Fighting*, 4th Edition, 1st Printing, page 576.
Jones and Bartlett, *Fundamentals of Fire Fighter Skills*, 1st Edition, 1st Printing, page 940.
Answer: B

91. Reference: NFPA 1001, 6.5.1 and 6.5.1(A)(B)
Delmar, *Firefighter's Handbook*, 2nd Edition, 1st Printing, page 324.
IFSTA, *Essentials of Fire Fighting*, 4th Edition, 1st Printing, page 577.
Jones and Bartlett, *Fundamentals of Fire Fighter Skills*, 1st Edition, 1st Printing, page 940
Answer: B

92. Reference: NFPA 1001, 6.5.1 and 6.5.1(A)
Delmar, *Firefighter's Handbook*, 2nd Edition, 1st Printing, page 324.
IFSTA, *Essentials of Fire Fighting*, 4th Edition, 1st Printing, page 577.
Jones and Bartlett, *Fundamentals of Fire Fighter Skills*, 1st Edition, 1st Printing, page 940.
Answer: D

93. Reference: NFPA 1001, 6.5.1 and 6.5.1(A)(B)
Delmar, *Firefighter's Handbook*, 2nd Edition, 1st Printing, page 324.
IFSTA, *Essentials of Fire Fighting*, 4th Edition, 1st Printing, page 577.
Jones and Bartlett, *Fundamentals of Fire Fighter Skills*, 1st Edition, 1st Printing, page 940.
Answer: B

94. Reference: NFPA 1001, 6.5.1 and 6.5.1(A)
Delmar, *Firefighter's Handbook*, 2nd Edition, 1st Printing, pages 318 and 323.
IFSTA, *Essentials of Fire Fighting*, 4th Edition, 1st Printing, pages 572-573.
Jones and Bartlett, *Fundamentals of Fire Fighter Skills*, 1st Edition, 1st Printing, page 939.
Answer: D

95. Reference: NFPA 1001, 6.5.1 and 6.5.1(A)(B)
Delmar, *Firefighter's Handbook*, 2nd Edition, 1st Printing, page 316.
IFSTA, *Essentials of Fire Fighting*, 4th Edition, 1st Printing, page 574.
Jones and Bartlett, *Fundamentals of Fire Fighter Skills*, 1st Edition, 1st Printing, page 936.
Answer: A

96. Reference: NFPA 1001, 6.5.4 and 6.5.4(A)(B)
Delmar, *Firefighter's Handbook*, 2nd Edition, 1st Printing, page 205.
IFSTA, *Essentials of Fire Fighting*, 4th Edition, 1st Printing, page 383.
Jones and Bartlett, *Fundamentals of Fire Fighter Skills*, 1st Edition, 1st Printing, page 448.
Answer: B

97. Reference: NFPA 1001, 6.5.4 and 6.5.4(A)(B)
Delmar, *Firefighter's Handbook*, 2nd Edition, 1st Printing, page 215.
IFSTA, *Essentials of Fire Fighting*, 4th Edition, 1st Printing, pages 386 and 389.
Jones and Bartlett, *Fundamentals of Fire Fighter Skills*, 1st Edition, 1st Printing, page 455.
Answer: D

98. Reference: NFPA 1001, 6.5.4 and 6.5.4(A)(B)
Delmar, *Firefighter's Handbook*, 2nd Edition, 1st Printing, page 215.
IFSTA, *Essentials of Fire Fighting*, 4th Edition, 1st Printing, page 389.
Jones and Bartlett, *Fundamentals of Fire Fighter Skills*, 1st Edition, 1st Printing, page 455.
Answer: B

99. Reference: NFPA 1001, 6.5.4 and 6.5.4(A)(B)
Delmar, *Firefighter's Handbook*, 2nd Edition, 1st Printing, page 215.
IFSTA, *Essentials of Fire Fighting*, 4th Edition, 1st Printing, pages 385-386.
Jones and Bartlett, *Fundamentals of Fire Fighter Skills*, 1st Edition, 1st Printing, pages 454-455.
Answer: D

100. Reference: NFPA 1001, 6.5.4 and 6.5.4(A)(B)
Delmar, *Firefighter's Handbook*, 2nd Edition, 1st Printing, page 215.
IFSTA, *Essentials of Fire Fighting*, 4th Edition, 1st Printing, page 394.
Jones and Bartlett, *Fundamentals of Fire Fighter Skills*, 1st Edition, 1st Printing, page 455.
Answer: B

Examination II-2 Answer Key

Directions
Follow these steps carefully for completing the feedback step of Systematic Approach to Exam Preparation (SAEP):

1. After entering your scores, look up the answers for the examination items you missed, as well as those you guessed, even if you guessed correctly. If you are guessing, it means the answer is not perfectly clear. In this process we are committed to making you as knowledgeable as possible.
2. Enter the number of missed and guessed examination items in the blank on the personal progress plotter.
3. Highlight the answer in the reference materials. Once you have highlighted the answer, read the paragraph preceding and the paragraph following the one in which the correct answer is located. Now that you have highlighted the answer, enter the paragraph number and page number next to the guessed or missed examination item on your examination. Count any part of a paragraph at the beginning of the page as one paragraph until you reach the paragraph containing your highlighted answer. This step will help you locate and review your missed and guessed examination items later in the process. This step is __essential__ to learning the material in context and by association. These learning techniques (context/association) are the very backbone of the SAEP approach.
4. Once you have completed the feedback step, you may proceed to the next examination.

1. Reference: NFPA 1001, 6.1.1.1 and 6.1.1.2
   Delmar, *Firefighter's Handbook*, 2nd Edition, 1st Printing, page 36.
   IFSTA, *Essentials of Fire Fighting*, 4th Edition, 1st Printing, page 16.
   Jones and Bartlett, *Fundamentals of Fire Fighter Skills*, 1st Edition, 1st Printing, pages 102 and 116.
   Answer: B

2. Reference: NFPA 1001, 6.1.1.1 and 6.1.1.2
   Delmar, *Firefighter's Handbook*, 2nd Edition, 1st Printing, page 725.
   IFSTA, *Essentials of Fire Fighting*, 4th Edition, 1st Printing, page 29.
   Jones and Bartlett, *Fundamentals of Fire Fighter Skills*, 1st Edition, 1st Printing, page 27.
   Answer: A

3. Reference: NFPA 1001, 6.1.1.1 and 6.1.1.2
   Delmar, *Firefighter's Handbook*, 2nd Edition, 1st Printing, page 725.
   IFSTA, *Essentials of Fire Fighting*, 4th Edition, 1st Printing, page 29.
   Jones and Bartlett, *Fundamentals of Fire Fighter Skills*, 1st Edition, 1st Printing, page 27.
   Answer: A

4. Reference: NFPA 1001, 6.1.1.1 and 6.1.1.2
Delmar, *Firefighter's Handbook*, 2nd Edition, 1st Printing, page 41.
IFSTA, *Essentials of Fire Fighting*, 4th Edition, 1st Printing, pages 14-15.
Jones and Bartlett, *Fundamentals of Fire Fighter Skills*, 1st Edition, 1st Printing, page 107.
Answer: D

5. Reference: NFPA 1001, 6.1.1.1 and 6.1.1.2
Delmar, *Firefighter's Handbook*, 2nd Edition, 1st Printing, page 38.
IFSTA, *Essentials of Fire Fighting*, 4th Edition, 1st Printing, page 14.
Jones and Bartlett, *Fundamentals of Fire Fighter Skills*, 1st Edition, 1st Printing, page 104.
Answer: C

6. Reference: NFPA 1001, 6.1.1.1 and 6.1.1.2
Delmar, *Firefighter's Handbook*, 2nd Edition, 1st Printing, page 39.
IFSTA, *Essentials of Fire Fighting*, 4th Edition, 1st Printing, page 15.
Jones and Bartlett, *Fundamentals of Fire Fighter Skills*, 1st Edition, 1st Printing, page 108.
Answer: B

7. Reference: NFPA 1001, 6.1.1.1 and 6.1.1.2
Delmar, *Firefighter's Handbook*, 2nd Edition, 1st Printing, page 41.
IFSTA, *Essentials of Fire Fighting*, 4th Edition, 1st Printing, page 14.
Jones and Bartlett, *Fundamentals of Fire Fighter Skills*, 1st Edition, 1st Printing, page 108.
Answer: D

8. Reference: NFPA 1001, 6.2.1 and 6.2.1(A)
IFSTA, *Essentials of Fire Fighting*, 4th Edition, 1st Printing, page 651.
Jones and Bartlett, *Fundamentals of Fire Fighter Skills*, 1st Edition, 1st Printing, page 89.
Answer: A

9. Reference: NFPA 1001, 6.2.1 and 6.2.1(A)
IFSTA, *Essentials of Fire Fighting*, 4th Edition, 1st Printing, page 651.
Jones and Bartlett, *Fundamentals of Fire Fighter Skills*, 1st Edition, 1st Printing, page 89.
Answer: D

10. Reference: NFPA 1001, 6.2.2 and 6.2.2(A)(B)
Delmar, *Firefighter's Handbook*, 2nd Edition, 1st Printing, page 68.
Jones and Bartlett, *Fundamentals of Fire Fighter Skills*, 1st Edition, 1st Printing, page 87.
Answer: C

11. Reference: NFPA 1001, 6.2.2 and 6.2.2(A)(B)
IFSTA, *Essentials of Fire Fighting*, 4th Edition, 1st Printing, page 647.
Jones and Bartlett, *Fundamentals of Fire Fighter Skills*, 1st Edition, 1st Printing, page 287.
Answer: C

12. Reference: NFPA 1001, 6.3.1 and 6.3.1(A)(B)
Delmar, *Firefighter's Handbook*, 2nd Edition, 1st Printing, page 297.
IFSTA, *Essentials of Fire Fighting*, 4th Edition, 1st Printing, pages 510-512.
Jones and Bartlett, *Fundamentals of Fire Fighter Skills*, 1st Edition, 1st Printing, page 527.
Answer: B

13. Reference: NFPA 1001, 6.3.1 and 6.3.1(A)(B)
Delmar, *Firefighter's Handbook*, 2nd Edition, 2nd Printing, page 295.
IFSTA, *Essentials of Fire Fighting*, 4th Edition, 1st Printing, page 499.
Jones and Bartlett, *Fundamentals of Fire Fighter Skills*, 1st Edition, 1st Printing, pages 527 and 529.
Answer: D

14. Reference: NFPA 1001, 6.3.1 and 6.3.1(A)(B)
Delmar, *Firefighter's Handbook*, 2nd Edition, 1st Printing, page 295.
IFSTA, *Essentials of Fire Fighting*, 4th Edition, 1st Printing, page 499.
Jones and Bartlett, *Fundamentals of Fire Fighter Skills*, 1st Edition, 1st Printing, page 529.
Answer: B

15. Reference: NFPA 1001, 6.3.1 and 6.3.1(A)(B)
Delmar, *Firefighter's Handbook*, 2nd Edition, 1st Printing, page 297.
IFSTA, *Essentials of Fire Fighting*, 4th Edition, 1st Printing, page 505.
Jones and Bartlett, *Fundamentals of Fire Fighter Skills*, 1st Edition, 1st Printing, page 525.
Answer: B

16. Reference: NFPA 1001, 6.3.1 and 6.3.1(A)(B)
Delmar, *Firefighter's Handbook*, 2nd Edition, 1st Printing, page 297.
IFSTA, *Essentials of Fire Fighting*, 4th Edition, 1st Printing, page 505.
Jones and Bartlett, *Fundamentals of Fire Fighter Skills*, 1st Edition, 1st Printing, page 526.
Answer: C

17. Reference: NFPA 1001, 6.3.1 and 6.3.1(A)(B)
Delmar, *Firefighter's Handbook*, 2nd Edition, 1st Printing, page 297.
IFSTA, *Essentials of Fire Fighting*, 4th Edition, 1st Printing, pages 501-503.
Jones and Bartlett, *Fundamentals of Fire Fighter Skills*, 1st Edition, 1st Printing, page 526.
Answer: A

18. Reference: NFPA 1001, 6.3.1 and 6.3.1(A)(B)
Delmar, *Firefighter's Handbook*, 2nd Edition, 1st Printing, page 296.
IFSTA, *Essentials of Fire Fighting*, 4th Edition, 1st Printing, page 507.
Jones and Bartlett, *Fundamentals of Fire Fighter Skills*, 1st Edition, 1st Printing, page 526.
Answer: D

19. Reference: NFPA 1001, 6.3.1 and 6.3.1(A)(B)
Delmar, *Firefighter's Handbook*, 2nd Edition, 1st Printing, page 624.
IFSTA, *Essentials of Fire Fighting*, 4th Edition, 1st Printing, pages 532-533.
Jones and Bartlett, *Fundamentals of Fire Fighter Skills*, 1st Edition, 1st Printing, page 640.
Answer: D

20. Reference: NFPA 1001, 6.3.1 and 6.3.1(A)(B)
Delmar, *Firefighter's Handbook*, 2nd Edition, 1st Printing, page 297.
IFSTA, *Essentials of Fire Fighting*, 4th Edition, 1st Printing, page 498.
Jones and Bartlett, *Fundamentals of Fire Fighter Skills*, 1st Edition, 1st Printing, page 187.
Answer: A

21. Reference: NFPA 1001, 6.3.2 and 6.3.2(A)(B)
Delmar, *Firefighter's Handbook*, 2nd Edition, 1st Printing, page 290.
IFSTA, *Essentials of Fire Fighting*, 4th Edition, 1st Printing, page 488.
Jones and Bartlett, *Fundamentals of Fire Fighter Skills*, 1st Edition, 1st Printing, page 793.
Answer: C

22. Reference: NFPA 1001, 6.3.2 and 6.3.2(A)(B)
Delmar, *Firefighter's Handbook*, 2nd Edition, 1st Printing, page 575.
IFSTA, *Essentials of Fire Fighting*, 4th Edition, 1st Printing, page 357.
Jones and Bartlett, *Fundamentals of Fire Fighter Skills*, 1st Edition, 1st Printing, page 414.
Answer: C

23. Reference: NFPA 1001, 6.3.2 and 6.3.2(A)(B)
Delmar, *Firefighter's Handbook*, 2nd Edition, 1st Printing, page 575.
IFSTA, *Essentials of Fire Fighting*, 4th Edition, 1st Printing, page 346.
Jones and Bartlett, *Fundamentals of Fire Fighter Skills*, 1st Edition, 1st Printing, page 423.
Answer: C

24. Reference: NFPA 1001, 6.3.2 and 6.3.2(A)(B)
Delmar, *Firefighter's Handbook*, 2nd Edition, 1st Printing, pages 575-576.
IFSTA, *Essentials of Fire Fighting*, 4th Edition, 1st Printing, page 354.
Jones and Bartlett, *Fundamentals of Fire Fighter Skills*, 1st Edition, 1st Printing, page 423.
Answer: B

25. Reference: NFPA 1001, 6.3.2 and 6.3.2(A)(B)
Delmar, *Firefighter's Handbook*, 2nd Edition, 1st Printing, page 553.
IFSTA, *Essentials of Fire Fighting*, 4th Edition, 1st Printing, page 367.
Jones and Bartlett, *Fundamentals of Fire Fighter Skills*, 1st Edition, 1st Printing, page 411.
Answer: C

26. Reference: NFPA 1001, 6.3.2 and 6.3.2(A)(B)
Delmar, *Firefighter's Handbook*, 2nd Edition, 1st Printing, page 575.
IFSTA, *Essentials of Fire Fighting*, 4th Edition, 1st Printing, pages 354 and 358.
Jones and Bartlett, *Fundamentals of Fire Fighter Skills*, 1st Edition, 1st Printing, page 429.
Answer: B

27. Reference: NFPA 1001, 6.3.2 and 6.3.2(A)(B)
Delmar, *Firefighter's Handbook*, 2nd Edition, 1st Printing, page 577.
IFSTA, *Essentials of Fire Fighting*, 4th Edition, 1st Printing, page 358.
Jones and Bartlett, *Fundamentals of Fire Fighter Skills*, 1st Edition, 1st Printing, page 431.
Answer: D

28. Reference: NFPA 1001, 6.3.2 and 6.3.2(A)(B)
Delmar, *Firefighter's Handbook*, 2nd Edition, 1st Printing, page 563.
IFSTA, *Essentials of Fire Fighting*, 4th Edition, 1st Printing, page 347.
Jones and Bartlett, *Fundamentals of Fire Fighter Skills*, 1st Edition, 1st Printing, page 419.
Answer: D

29. Reference: NFPA 1001, 6.3.2 and 6.3.2(A)(B)
Delmar, *Firefighter's Handbook*, 2nd Edition, 1st Printing, page 650.
IFSTA, *Essentials of Fire Fighting*, 4th Edition, 1st Printing, page 597.
Jones and Bartlett, *Fundamentals of Fire Fighter Skills*, 1st Edition, 1st Printing, page 574.
Answer: B

30. Reference: NFPA 1001, 6.3.2 and 6.3.2(A)(B)
Delmar, *Firefighter's Handbook*, 2nd Edition, 1st Printing, page 565.
IFSTA, *Essentials of Fire Fighting*, 4th Edition, 1st Printing, page 488.
Jones and Bartlett, *Fundamentals of Fire Fighter Skills*, 1st Edition, 1st Printing, page 521.
Answer: C

31. Reference: NFPA 1001, 6.3.2 and 6.3.2(A)(B)
Delmar, *Firefighter's Handbook*, 2nd Edition, 1st Printing, page 283.
IFSTA, *Essentials of Fire Fighting*, 4th Edition, 1st Printing, page 494.
Jones and Bartlett, *Fundamentals of Fire Fighter Skills*, 1st Edition, 1st Printing, page 520.
Answer: B

32. Reference: NFPA 1001, 6.3.2 and 6.3.2(A)(B)
Delmar, *Firefighter's Handbook*, 2nd Edition, 1st Printing, page 348.
IFSTA, *Essentials of Fire Fighting*, 4th Edition, 1st Printing, pages 67-68.
Jones and Bartlett, *Fundamentals of Fire Fighter Skills*, 1st Edition, 1st Printing, page 159.
Answer: B

33. Reference: NFPA 1001, 6.3.2 and 6.3.2(A)(B)
Delmar, *Firefighter's Handbook*, 2nd Edition, 1st Printing, page 205.
IFSTA, *Essentials of Fire Fighting*, 4th Edition, 1st Printing, page 391.
Jones and Bartlett, *Fundamentals of Fire Fighter Skills*, 1st Edition, 1st Printing, page 458.
Answer: C

34. Reference: NFPA 1001, 6.3.2 and 6.3.2(A)(B)
Delmar, *Firefighter's Handbook*, 2nd Edition, 1st Printing, pages 210-211.
IFSTA, *Essentials of Fire Fighting*, 4th Edition, 1st Printing, page 392.
Jones and Bartlett, *Fundamentals of Fire Fighter Skills*, 1st Edition, 1st Printing, page 459.
Answer: C

35. Reference: NFPA 1001, 6.3.2 and 6.3.2(A)(B)
Delmar, *Firefighter's Handbook*, 2nd Edition, 1st Printing, page 617.
IFSTA, *Essentials of Fire Fighting*, 4th Edition, 1st Printing, page 545.
Jones and Bartlett, *Fundamentals of Fire Fighter Skills*, 1st Edition, 1st Printing, page 636.
Answer: B

36. Reference: NFPA 1001, 6.3.2 and 6.3.2(A)(B)
Delmar, *Firefighter's Handbook*, 2nd Edition, 1st Printing, pages 356-357.
IFSTA, *Essentials of Fire Fighting*, 4th Edition, 1st Printing, page 361.
Jones and Bartlett, *Fundamentals of Fire Fighter Skills*, 1st Edition, 1st Printing, page 158.
Answer: C

37. Reference: NFPA 1001, 6.3.2 and 6.3.2(A)(B)
Delmar, *Firefighter's Handbook*, 2nd Edition, 1st Printing, page 462.
IFSTA, *Essentials of Fire Fighting*, 4th Edition, 1st Printing, page 175.
Jones and Bartlett, *Fundamentals of Fire Fighter Skills*, 1st Edition, 1st Printing, page 375.
Answer: D

38. Reference: NFPA 1001, 6.3.2 and 6.3.2(A)(B)
Delmar, *Firefighter's Handbook*, 2nd Edition, 1st Printing, page 575.
IFSTA, *Essentials of Fire Fighting*, 4th Edition, 1st Printing, page 346.
Jones and Bartlett, *Fundamentals of Fire Fighter Skills*, 1st Edition, 1st Printing, page 423.
Answer: A

39. Reference: NFPA 1001, 6.3.2 and 6.3.2(A)(B)
Delmar, *Firefighter's Handbook*, 2nd Edition, 1st Printing, page 594.
IFSTA, *Essentials of Fire Fighting*, 4th Edition, 1st Printing, page 73.
Jones and Bartlett, *Fundamentals of Fire Fighter Skills*, 1st Edition, 1st Printing, page 290.
Answer: D

40. Reference: NFPA 1001, 6.3.2 and 6.3.2(A)(B)
Delmar, *Firefighter's Handbook*, 2nd Edition, 1st Printing, page 607.
IFSTA, *Essentials of Fire Fighting*, 4th Edition, 1st Printing, page 541.
Jones and Bartlett, *Fundamentals of Fire Fighter Skills*, 1st Edition, 1st Printing, page 291.
Answer: B

41. Reference: NFPA 1001, 6.3.2 and 6.3.2(A)(B)
Delmar, *Firefighter's Handbook*, 2nd Edition, 1st Printing, page 462.
IFSTA, *Essentials of Fire Fighting*, 4th Edition, 1st Printing, page 522.
Jones and Bartlett, *Fundamentals of Fire Fighter Skills*, 1st Edition, 1st Printing, page 27.
Answer: A

42. Reference: NFPA 1001, 6.3.2 and 6.3.2(A)(B)
Delmar, *Firefighter's Handbook*, 2nd Edition, 1st Printing, pages 618-619.
IFSTA, *Essentials of Fire Fighting*, 4th Edition, 1st Printing, page 545.
Jones and Bartlett, *Fundamentals of Fire Fighter Skills*, 1st Edition, 1st Printing, page 437.
Answer: C

43. Reference: NFPA 1001, 6.3.2 and 6.3.2(A)(B)
Delmar, *Firefighter's Handbook*, 2nd Edition, 1st Printing, page 465.
IFSTA, *Essentials of Fire Fighting*, 4th Edition, 1st Printing, pages 176-178.
Jones and Bartlett, *Fundamentals of Fire Fighter Skills*, 1st Edition, 1st Printing, page 378.
Answer: A

44. Reference: NFPA 1001, 6.3.2 and 6.3.2(A)(B)
Delmar, *Firefighter's Handbook*, 2nd Edition, 1st Printing, page 609.
IFSTA, *Essentials of Fire Fighting*, 4th Edition, 1st Printing, page 522.
Jones and Bartlett, *Fundamentals of Fire Fighter Skills*, 1st Edition, 1st Printing, page 374.
Answer: A

45. Reference: NFPA 1001, 6.3.2 and 6.3.2(A)(B)
Delmar, *Firefighter's Handbook*, 2nd Edition, 1st Printing, page 579.
IFSTA, *Essentials of Fire Fighting*, 4th Edition, 1st Printing, pages 356 and 361.
Jones and Bartlett, *Fundamentals of Fire Fighter Skills*, 1st Edition, 1st Printing, pages 434-435.
Answer: A

46. Reference: NFPA 1001, 6.3.2 and 6.3.2(A)(B)
Delmar, *Firefighter's Handbook*, 2nd Edition, 1st Printing, page 88.
IFSTA, *Essentials of Fire Fighting*, 4th Edition, 1st Printing, page 534.
Jones and Bartlett, *Fundamentals of Fire Fighter Skills*, 1st Edition, 1st Printing, page 640.
Answer: C

47. Reference: NFPA 1001, 6.3.2 and 6.3.2(A)(B)
Delmar, *Firefighter's Handbook*, 2nd Edition, 1st Printing, page 147.
IFSTA, *Essentials of Fire Fighting*, 4th Edition, 1st Printing, page 596.
Jones and Bartlett, *Fundamentals of Fire Fighter Skills*, 1st Edition, 1st Printing, page 574.
Answer: C

48. Reference: NFPA 1001, 6.3.2 and 6.3.2(A)(B)
Delmar, *Firefighter's Handbook*, 2nd Edition, 1st Printing, page 609.
IFSTA, *Essentials of Fire Fighting*, 4th Edition, 1st Printing, page 522.
Jones and Bartlett, *Fundamentals of Fire Fighter Skills*, 1st Edition, 1st Printing, page 374.
Answer: A

49. Reference: NFPA 1001, 6.3.3 and 6.3.3(A)(B)
Delmar, *Firefighter's Handbook*, 2nd Edition, 1st Printing, page 866.
IFSTA, *Essentials of Fire Fighting*, 4th Edition, 1st Printing, page 530.
Jones and Bartlett, *Fundamentals of Fire Fighter Skills*, 1st Edition, 1st Printing, page 641.
Answer: B

50. Reference: NFPA 1001, 6.3.3 and 6.3.3(A)(B)
Delmar, *Firefighter's Handbook*, 2nd Edition, 1st Printing, page 866.
IFSTA, *Essentials of Fire Fighting*, 4th Edition, 1st Printing, page 530.
Jones and Bartlett, *Fundamentals of Fire Fighter Skills*, 1st Edition, 1st Printing, page 641.
Answer: B

51. Reference: NFPA 1001, 6.3.3 and 6.3.3(A)(B)
Delmar, *Firefighter's Handbook*, 2nd Edition, 1st Printing, pages 128-130.
IFSTA, *Essentials of Fire Fighting*, 4th Edition, 1st Printing, pages 82-83, 534.
Jones and Bartlett, *Fundamentals of Fire Fighter Skills*, 1st Edition, 1st Printing, pages 640-641.
Answer: C

52. Reference: NFPA 1001, 6.3.4 and 6.3.4(A)(B)
Delmar, *Firefighter's Handbook*, 1st Edition, 1st Printing, page 654.
Jones and Bartlett, *Fundamentals of Fire Fighter Skills*, 1st Edition, 1st Printing, pages 965-966.
Answer: D

53. Reference: NFPA 1001, 6.3.4 and 6.3.4(A)(B)
Delmar, *Firefighter's Handbook*, 2nd Edition, 1st Printing, page 654.
Jones and Bartlett, *Fundamentals of Fire Fighter Skills*, 1st Edition, 1st Printing, page 966.
Answer: C

54. Reference: NFPA 1001, 6.3.2 and 6.3.2(A)(B)
Delmar, *Firefighter's Handbook*, 1st Edition, 1st Printing, page 654.
Jones and Bartlett, *Fundamentals of Fire Fighter Skills*, 1st Edition, 1st Printing, page 965.
Answer: D

55. Reference: NFPA 1001, 6.3.4 and 6.3.4(A)(B)
IFSTA, *Essentials of Fire Fighting*, 4th Edition, 1st Printing, pages 625-626.
Jones and Bartlett, *Fundamentals of Fire Fighter Skills*, 1st Edition, 1st Printing, pages 972-973.
Answer: C

56. Reference: NFPA 1001, 6.4.1, 6.4.1(A)(B), 6.4.2, and 6.4.2(A)(B)
Delmar, *Firefighter's Handbook*, 2nd Edition, 1st Printing, page 478.
IFSTA, *Essentials of Fire Fighting*, 4th Edition, 1st Printing, page 185.
Jones and Bartlett, *Fundamentals of Fire Fighter Skills*, 1st Edition, 1st Printing, page 396.
Answer: C

57. Reference: NFPA 1001, 6.4.1 and 6.4.1(A)
Delmar, *Firefighter's Handbook*, 2nd Edition, 1st Printing, page 542.
IFSTA, *Essentials of Fire Fighting*, 4th Edition, 1st Printing, page 203.
Jones and Bartlett, *Fundamentals of Fire Fighter Skills*, 1st Edition, 1st Printing, page 741.
Answer: A

58. Reference: NFPA 1001, 6.4.1 and 6.4.1(A)
Delmar, *Firefighter's Handbook*, 1st Edition, 1st Printing, page 542.
IFSTA, *Essentials of Fire Fighting*, 4th Edition, 1st Printing, page 202.
Jones and Bartlett, *Fundamentals of Fire Fighter Skills*, 1st Edition, 1st Printing, page 741.
Answer: B

59. Reference: NFPA 1001, 6.4.1 and 6.4.1(A)(B)
Delmar, *Firefighter's Handbook*, 2nd Edition, 1st Printing, page 488.
IFSTA, *Essentials of Fire Fighting*, 4th Edition, 1st Printing, page 197.
Jones and Bartlett, *Fundamentals of Fire Fighter Skills*, 1st Edition, 1st Printing, page 736.
Answer: B

60. Reference: NFPA 1001, 6.4.1 and 6.4.1(A)(B)
Delmar, *Firefighter's Handbook*, 2nd Edition, 1st Printing, page 488.
IFSTA, *Essentials of Fire Fighting*, 4th Edition, 1st Printing, page 202.
Jones and Bartlett, *Fundamentals of Fire Fighter Skills*, 1st Edition, 1st Printing, page 742.
Answer: C

61. Reference: NFPA 1001, 6.4.1 and 6.4.1(A)(B)
Delmar, *Firefighter's Handbook*, 2nd Edition, 1st Printing, page 488.
IFSTA, *Essentials of Fire Fighting*, 4th Edition, 1st Printing, page 202.
Jones and Bartlett, *Fundamentals of Fire Fighter Skills*, 1st Edition, 1st Printing, page 746.
Answer: B

62. Reference: NFPA 1001, 6.4.1, 6.4.1(A)(B), 6.4.2, and 6.4.2(A)(B)
Delmar, *Firefighter's Handbook*, 2nd Edition, 1st Printing, page 484.
IFSTA, *Essentials of Fire Fighting*, 4th Edition, 1st Printing, page 196.
Jones and Bartlett, *Fundamentals of Fire Fighter Skills*, 1st Edition, 1st Printing, pages 738-739.
Answer: D

63. Reference: NFPA 1001, 6.4.1 and 6.4.1(A)(B)
Delmar, *Firefighter's Handbook*, 2nd Edition, 1st Printing, page 489.
IFSTA, *Essentials of Fire Fighting*, 4th Edition, 1st Printing, pages 201-202.
Jones and Bartlett, *Fundamentals of Fire Fighter Skills*, 1st Edition, 1st Printing, page 740.
Answer: A

64. Reference: NFPA 1001, 6.4.2 and 6.4.2(A)(B)
Delmar, *Firefighter's Handbook*, 2nd Edition, 1st Printing, page 467.
IFSTA, *Essentials of Fire Fighting*, 4th Edition, 1st Printing, page 184.
Jones and Bartlett, *Fundamentals of Fire Fighter Skills*, 1st Edition, 1st Printing, page 28.
Answer: B

65. Reference: NFPA 1001, 6.4.2 and 6.4.2(A)(B)
Delmar, *Firefighter's Handbook*, 2nd Edition, 1st Printing, page 469.
IFSTA, *Essentials of Fire Fighting*, 4th Edition, 1st Printing, pages 228-229.
Jones and Bartlett, *Fundamentals of Fire Fighter Skills*, 1st Edition, 1st Printing, page 392.
Answer: B

66. Reference: NFPA 1001, 6.4.2 and 6.4.2(A)(B)
Delmar, *Firefighter's Handbook*, 2nd Edition, 1st Printing, page 492.
IFSTA, *Essentials of Fire Fighting*, 4th Edition, 1st Printing, page 212.
Jones and Bartlett, *Fundamentals of Fire Fighter Skills*, 1st Edition, 1st Printing, pages 767-768.
Answer: A

67. Reference: NFPA 1001, 6.4.2 and 6.4.2(A)(B)
Delmar, *Firefighter's Handbook*, 2nd Edition, 1st Printing, pages 732-733.
IFSTA, *Essentials of Fire Fighting*, 4th Edition, 1st Printing, page 544.
Jones and Bartlett, *Fundamentals of Fire Fighter Skills*, 1st Edition, 1st Printing, page 27.
Answer: C

68. Reference: NFPA 1001, 6.5.1 and 6.5.1(A)(B)
Delmar, *Firefighter's Handbook*, 2nd Edition, 1st Printing, page 323.
IFSTA, *Essentials of Fire Fighting*, 4th Edition, 1st Printing, page 579.
Jones and Bartlett, *Fundamentals of Fire Fighter Skills*, 1st Edition, 1st Printing, pages 940-941.
Answer: B

69. Reference: NFPA 1001, 6.5.1 and 6.5.1(A)(B)
Delmar, *Firefighter's Handbook*, 2nd Edition, 1st Printing, page 314.
IFSTA, *Essentials of Fire Fighting*, 4th Edition, 1st Printing, page 572.
Jones and Bartlett, *Fundamentals of Fire Fighter Skills*, 1st Edition, 1st Printing, page 935.
Answer: C

70. Reference: NFPA 1001, 6.5.1 and 6.5.1(A)(B)
Delmar, *Firefighter's Handbook*, 2nd Edition, 1st Printing, pages 315 and 317.
IFSTA, *Essentials of Fire Fighting*, 4th Edition, 1st Printing, pages 574-575.
Jones and Bartlett, *Fundamentals of Fire Fighter Skills*, 1st Edition, 1st Printing, page 938.
Answer: D

71. Reference: NFPA 1001, 6.5.1 and 6.5.1(A)(B)
Delmar, *Firefighter's Handbook*, 2nd Edition, 1st Printing, page 318.
IFSTA, *Essentials of Fire Fighting*, 4th Edition, 1st Printing, page 581.
Jones and Bartlett, *Fundamentals of Fire Fighter Skills*, 1st Edition, 1st Printing, page 944.
Answer: C

72. Reference: NFPA 1001, 6.5.1 and 6.5.1(A)(B)
Delmar, *Firefighter's Handbook*, 2nd Edition, 1st Printing, page 318.
IFSTA, *Essentials of Fire Fighting*, 4th Edition, 1st Printing, page 581.
Jones and Bartlett, *Fundamentals of Fire Fighter Skills*, 1st Edition, 1st Printing, pages 944-945.
Answer: B

73. Reference: NFPA 1001, 6.5.1 and 6.5.1(A)(B)
Delmar, *Firefighter's Handbook*, 2nd Edition, 1st Printing, page 685.
IFSTA, *Essentials of Fire Fighting*, 4th Edition, 1st Printing, page 664.
Jones and Bartlett, *Fundamentals of Fire Fighter Skills*, 1st Edition, 1st Printing, page 651.
Answer: B

74. Reference: NFPA 1001, 6.5.1 and 6.5.1(A)
Delmar, *Firefighter's Handbook*, 2nd Edition, 1st Printing, page 685.
IFSTA, *Essentials of Fire Fighting*, 4th Edition, 1st Printing, page 664.
Jones and Bartlett, *Fundamentals of Fire Fighter Skills*, 1st Edition, 1st Printing, page 652.
Answer: A

75. Reference: NFPA 1001, 6.5.1 and 6.5.1(A)(B)
Delmar, *Firefighter's Handbook*, 2nd Edition, 1st Printing, pages 678-679.
IFSTA, *Essentials of Fire Fighting*, 4th Edition, 1st Printing, page 673.
Jones and Bartlett, *Fundamentals of Fire Fighter Skills*, 1st Edition, 1st Printing, page 923.
Answer: C

76. Reference: NFPA 1001, 6.5.1 and 6.5.1(A)(B)
Delmar, *Firefighter's Handbook*, 2nd Edition, 1st Printing, page 661.
IFSTA, *Essentials of Fire Fighting*, 4th Edition, 1st Printing, page 661.
Jones and Bartlett, *Fundamentals of Fire Fighter Skills*, 1st Edition, 1st Printing, page 651.
Answer: B

77. Reference: NFPA 1001, 6.5.1 and 6.5.1(A)(B)
Delmar, *Firefighter's Handbook*, 2nd Edition, 1st Printing, page 205.
IFSTA, *Essentials of Fire Fighting*, 4th Edition, 1st Printing, page 380.
Jones and Bartlett, *Fundamentals of Fire Fighter Skills*, 1st Edition, 1st Printing, page 448.
Answer: C

78. Reference: NFPA 1001, 6.5.1 and 6.5.1(A)(B)
Delmar, *Firefighter's Handbook*, 2nd Edition, 1st Printing, page 205.
IFSTA, *Essentials of Fire Fighting*, 4th Edition, 1st Printing, pages 382-383.
Jones and Bartlett, *Fundamentals of Fire Fighter Skills*, 1st Edition, 1st Printing, page 448.
Answer: D

79. Reference: NFPA 1001, 6.5.1 and 6.5.1(A)
Delmar, *Firefighter's Handbook*, 2nd Edition, 1st Printing, pages 215-216.
IFSTA, *Essentials of Fire Fighting*, 4th Edition, 1st Printing, page 386.
Jones and Bartlett, *Fundamentals of Fire Fighter Skills*, 1st Edition, 1st Printing, page 455.
Answer: B

80. Reference: NFPA 1001, 6.5.1 and 6.5.1(A)
Delmar, *Firefighter's Handbook*, 2nd Edition, 1st Printing, pages 215-216.
IFSTA, *Essentials of Fire Fighting*, 4th Edition, 1st Printing, pages 386 and 389.
Jones and Bartlett, *Fundamentals of Fire Fighter Skills*, 1st Edition, 1st Printing, page 455.
Answer: A

81. Reference: NFPA 1001, 6.5.1 and 6.5.1(A)
IFSTA, *Essentials of Fire Fighting*, 4th Edition, 1st Printing, page 386.
Jones and Bartlett, *Fundamentals of Fire Fighter Skills*, 1st Edition, 1st Printing, page 455.
Answer: A

82. Reference: NFPA 1001, 6.5.1 and 6.5.1(A)(B)
Delmar, *Firefighter's Handbook*, 2nd Edition, 1st Printing, page 292.
IFSTA, *Essentials of Fire Fighting*, 4th Edition, 1st Printing, page 490.
Jones and Bartlett, *Fundamentals of Fire Fighter Skills*, 1st Edition, 1st Printing, page 466.
Answer: C

83. Reference: NFPA 1001, 6.5.1 and 6.5.1(A)(B)
Delmar, *Firefighter's Handbook*, 2nd Edition, 1st Printing, page 207.
IFSTA, *Essentials of Fire Fighting*, 4th Edition, 1st Printing, page 388.
Jones and Bartlett, *Fundamentals of Fire Fighter Skills*, 1st Edition, 1st Printing, page 450.
Answer: B

84. Reference: NFPA 1001, 6.5.1 and 6.5.1(A)(B)
Delmar, *Firefighter's Handbook*, 2nd Edition, 1st Printing, page 310.
IFSTA, *Essentials of Fire Fighting*, 4th Edition, 1st Printing, page 562.
Jones and Bartlett, *Fundamentals of Fire Fighter Skills*, 1st Edition, 1st Printing, page 929.
Answer: C

85. Reference: NFPA 1001, 6.5.1 and 6.5.1(A)
Delmar, *Firefighter's Handbook*, 2nd Edition, 1st Printing, page 320.
IFSTA, *Essentials of Fire Fighting*, 4th Edition, 1st Printing, pages 581-582.
Jones and Bartlett, *Fundamentals of Fire Fighter Skills*, 1st Edition, 1st Printing, page 946.
Answer: C

86. Reference: NFPA 1001, 6.5.1 and 6.5.1(A)(B)
Delmar, *Firefighter's Handbook*, 2nd Edition, 1st Printing, page 321.
IFSTA, *Essentials of Fire Fighting*, 4th Edition, 1st Printing, pages 582-583.
Jones and Bartlett, *Fundamentals of Fire Fighter Skills*, 1st Edition, 1st Printing, page 947.
Answer: A

87. Reference: NFPA 1001, 6.5.1 and 6.5.1(A)
Delmar, *Firefighter's Handbook*, 2nd Edition, 1st Printing, page 354.
IFSTA, *Essentials of Fire Fighting*, 4th Edition, 1st Printing, page 66.
Jones and Bartlett, *Fundamentals of Fire Fighter Skills*, 1st Edition, 1st Printing, page 149.
Answer: A

88. Reference: NFPA 1001, 6.5.1 and 6.5.1(A)
Delmar, *Firefighter's Handbook*, 2nd Edition, 1st Printing, page 361.
IFSTA, *Essentials of Fire Fighting*, 4th Edition, 1st Printing, page 75.
Jones and Bartlett, *Fundamentals of Fire Fighter Skills*, 1st Edition, 1st Printing, page 427.
Answer: B

89. Reference: NFPA 1001, 6.5.1 and 6.5.1(A)(B)
Delmar, *Firefighter's Handbook*, 2nd Edition, 1st Printing, page 329.
IFSTA, *Essentials of Fire Fighting*, 4th Edition, 1st Printing, pages 583-584.
Jones and Bartlett, *Fundamentals of Fire Fighter Skills*, 1st Edition, 1st Printing, page 946.
Answer: B

90. Reference: NFPA 1001, 6.5.1 and 6.5.1(A)(B)
Delmar, *Firefighter's Handbook*, 2nd Edition, 1st Printing, page 325.
IFSTA, *Essentials of Fire Fighting*, 4th Edition, 1st Printing, page 577.
Jones and Bartlett, *Fundamentals of Fire Fighter Skills*, 1st Edition, 1st Printing, page 940.
Answer: C

91. Reference: NFPA 1001, 6.5.2 and 6.5.2(A)(B)
IFSTA, *Essentials of Fire Fighting*, 4th Edition, 1st Printing, page 188.
Jones and Bartlett, *Fundamentals of Fire Fighter Skills*, 1st Edition, 1st Printing, page 583.
Answer: D

92. Reference: NFPA 1001, 6.5.2 and 6.5.2(A)(B)
IFSTA, *Essentials of Fire Fighting*, 4th Edition, 1st Printing, page 188.
Jones and Bartlett, *Fundamentals of Fire Fighter Skills*, 1st Edition, 1st Printing, page 581.
Answer: B

93. Reference: NFPA 1001, 6.5.2 and 6.5.2(A)(B)
IFSTA, *Essentials of Fire Fighting*, 4th Edition, 1st Printing, page 187.
Jones and Bartlett, *Fundamentals of Fire Fighter Skills*, 1st Edition, 1st Printing, page 580.
Answer: C

94. Reference: NFPA 1001, 6.5.2 and 6.5.2(A)(B)
IFSTA, *Essentials of Fire Fighting*, 4th Edition, 1st Printing, page 187.
Jones and Bartlett, *Fundamentals of Fire Fighter Skills*, 1st Edition, 1st Printing, page 581.
Answer: D

95. Reference: NFPA 1001, 6.5.3 and 6.5.3(A)(B)
Delmar, *Firefighter's Handbook*, 2nd Edition, 1st Printing, page 274.
IFSTA, *Essentials of Fire Fighting*, 4th Edition, 1st Printing, pages 437 and 483.
Jones and Bartlett, *Fundamentals of Fire Fighter Skills*, 1st Edition, 1st Printing, page 479.
Answer: C

96. Reference: NFPA 1001, 6.5.3 and 6.5.3(A)(B)
Delmar, *Firefighter's Handbook*, 2nd Edition, 1st Printing, page 274.
IFSTA, *Essentials of Fire Fighting*, 4th Edition, 1st Printing, pages 437 and 483.
Jones and Bartlett, *Fundamentals of Fire Fighter Skills*, 1st Edition, 1st Printing, page 479.
Answer: D

97. Reference: NFPA 1001, 6.5.3 and 6.5.3(A)(B)
Delmar, *Firefighter's Handbook*, 2nd Edition, 1st Printing, page 274.
IFSTA, *Essentials of Fire Fighting*, 4th Edition, 1st Printing, page 437.
Answer: B

98. Reference: NFPA 1001, 6.5.4 and 6.5.4(A)(B)
Delmar, *Firefighter's Handbook*, 2nd Edition, 1st Printing, page 208.
IFSTA, *Essentials of Fire Fighting*, 4th Edition, 1st Printing, page 383.
Jones and Bartlett, *Fundamentals of Fire Fighter Skills*, 1st Edition, 1st Printing, page 449.
Answer: A

99. Reference: NFPA 1001, 5.3.15 and 5.3.15(A)(B)
Delmar, *Firefighter's Handbook*, 2nd Edition, 1st Printing, page 208.
IFSTA, *Essentials of Fire Fighting*, 4th Edition, 1st Printing, page 390.
Jones and Bartlett, *Fundamentals of Fire Fighter Skills*, 1st Edition, 1st Printing, page 458.
Answer: B

100. Reference: NFPA 1001, 6.5.4 and 6.5.4(A)(B)
Delmar, *Firefighter's Handbook*, 2nd Edition, 1st Printing, page 207.
IFSTA, *Essentials of Fire Fighting*, 4th Edition, 1st Printing, page 387.
Jones and Bartlett, *Fundamentals of Fire Fighter Skills*, 1st Edition, 1st Printing, page 450.
Answer: D

Examination II-3 Answer Key

Directions
Follow these steps carefully for completing the feedback step of Systematic Approach to Exam Preparation (SAEP):

1. After entering your scores, look up the answers for the examination items you missed, as well as those you guessed, even if you guessed correctly. If you are guessing, it means the answer is not perfectly clear. In this process we are committed to making you as knowledgeable as possible.
2. Enter the number of missed and guessed examination items in the blank on the personal progress plotter.
3. Highlight the answer in the reference materials. Once you have highlighted the answer, read the paragraph preceding and the paragraph following the one in which the correct answer is located. Now that you have highlighted the answer, enter the paragraph number and page number next to the guessed or missed examination item on your examination. Count any part of a paragraph at the beginning of the page as one paragraph until you reach the paragraph containing your highlighted answer. This step will help you locate and review your missed and guessed examination items later in the process. This step is **essential** to learning the material in context and by association. These learning techniques (context/association) are the very backbone of the SAEP approach.
4. **Congratulations!** You have completed the Examination and feedback steps of SAEP when you have highlighted your guessed and missed examination items for this examination.

Proceed to Phase III and Phase IV. Study the materials carefully in these important phases. They will help you polish your examination taking skills. Approximately two to three days **prior** to taking your next examination, carefully read all the highlighted information in the reference materials using the same techniques applied during the feedback step. This will reinforce your learning and provide you with an added level of confidence going into the examination.

Someone once said to professional golfer, Tom Watson, after he won several tournament championships, "You are really lucky to have won those championships. You are really on a streak." Tom was reported to have said, "Yes, there is some luck involved, but what I have really noticed is that the more I practice the luckier I get." What Watson was saying is that good luck usually is the result of good preparation. This line of thinking certainly applies to learning the rules and hints of examination taking.

──────── Rule 7 ────────
Good luck = Good Preparation.

1. Reference: NFPA 1001, 6.1.1.1 and 6.1.1.2
   Delmar, *Firefighter's Handbook*, 2nd Edition, 1st Printing, page 36.
   IFSTA, *Essentials of Fire Fighting*, 4th Edition, 1st Printing, pages 541 and 544.
   Jones and Bartlett, *Fundamentals of Fire Fighter Skills*, 1st Edition, 1st Printing, page 117.
   Answer: B

2. Reference: NFPA 1001, 6.1.1.1 and 6.1.1.2
Delmar, *Firefighter's Handbook*, 2nd Edition, 1st Printing, page 36.
IFSTA, *Essentials of Fire Fighting*, 4th Edition, 1st Printing, page 7.
Jones and Bartlett, *Fundamentals of Fire Fighter Skills*, 1st Edition, 1st Printing, pages 8, 104.
Answer: D

3. Reference: NFPA 1001, 6.1.1.1 and 6.1.1.2
Delmar, *Firefighter's Handbook*, 2nd Edition, 1st Printing, page 39.
IFSTA, *Essentials of Fire Fighting*, 4th Edition, 1st Printing, page 14.
Jones and Bartlett, *Fundamentals of Fire Fighter Skills*, 1st Edition, 1st Printing, page 108.
Answer: B

4. Reference: NFPA 1001, 6.1.1.1 and 6.1.1.2
Delmar, *Firefighter's Handbook*, 2nd Edition, 1st Printing, page 39.
IFSTA, *Essentials of Fire Fighting*, 4th Edition, 1st Printing, page 14.
Jones and Bartlett, *Fundamentals of Fire Fighter Skills*, 1st Edition, 1st Printing, page 108.
Answer: D

5. Reference: NFPA 1001, 6.1.1.1 and 6.1.1.2
Delmar, *Firefighter's Handbook*, 2nd Edition, 1st Printing, page 41.
IFSTA, *Essentials of Fire Fighting*, 4th Edition, 1st Printing, page 15.
Jones and Bartlett, *Fundamentals of Fire Fighter Skills*, 1st Edition, 1st Printing, page 111.
Answer: C

6. Reference: NFPA 1001, 6.1.1.1 and 6.1.1.2
Delmar, *Firefighter's Handbook*, 2nd Edition, 1st Printing, page 41.
IFSTA, *Essentials of Fire Fighting*, 4th Edition, 1st Printing, pages 14-15.
Jones and Bartlett, *Fundamentals of Fire Fighter Skills*, 1st Edition, 1st Printing, page 107.
Answer: D

7. Reference: NFPA 1001, 6.1.1.1 and 6.1.1.2
Delmar, *Firefighter's Handbook*, 2nd Edition, 1st Printing, page 41.
IFSTA, *Essentials of Fire Fighting*, 4th Edition, 1st Printing, page 14.
Jones and Bartlett, *Fundamentals of Fire Fighter Skills*, 1st Edition, 1st Printing, page 108.
Answer: D

8. Reference: NFPA 1001, 6.1.1.1 and 6.1.1.2
Delmar, *Firefighter's Handbook*, 2nd Edition, 1st Printing, page 68.
IFSTA, *Essentials of Fire Fighting*, 4th Edition, 1st Printing, page 648.
Jones and Bartlett, *Fundamentals of Fire Fighter Skills*, 1st Edition, 1st Printing, page 87.
Answer: D

9. Reference: NFPA 1001, 6.2.1 and 6.2.1(A)(B)
   Delmar, *Firefighter's Handbook*, 2nd Edition, 1st Printing, page 69.
   IFSTA, *Essentials of Fire Fighting*, 4th Edition, 1st Printing, page 651.
   Jones and Bartlett, *Fundamentals of Fire Fighter Skills*, 1st Edition, 1st Printing, page 89.
   Answer: D

10. Reference: NFPA 1001, 6.2.1 and 6.2.1(A)(B)
    Delmar, *Firefighter's Handbook*, 2nd Edition, 1st Printing, page 69.
    IFSTA, *Essentials of Fire Fighting*, 4th Edition, 1st Printing, page 651.
    Jones and Bartlett, *Fundamentals of Fire Fighter Skills*, 1st Edition, 1st Printing, page 89.
    Answer: A

11. Reference: NFPA 1001, 6.2.2 and 6.2.2(A)(B)
    Delmar, *Firefighter's Handbook*, 2nd Edition, 1st Printing, page 63.
    IFSTA, *Essentials of Fire Fighting*, 4th Edition, 1st Printing, page 645.
    Jones and Bartlett, *Fundamentals of Fire Fighter Skills*, 1st Edition, 1st Printing, page 83.
    Answer: C

12. Reference: NFPA 1001, 6.2.2 and 6.2.2(A)(B)
    Delmar, *Firefighter's Handbook*, 2nd Edition, 1st Printing, page 51.
    IFSTA, *Essentials of Fire Fighting*, 4th Edition, 1st Printing, page 638.
    Jones and Bartlett, *Fundamentals of Fire Fighter Skills*, 1st Edition, 1st Printing, page 74.
    Answer: D

13. Reference: NFPA 1001, 6.2.2 and 6.2.2(A)(B)
    Delmar, *Firefighter's Handbook*, 2nd Edition, 1st Printing, page 67.
    IFSTA, *Essentials of Fire Fighting*, 4th Edition, 1st Printing, page 648.
    Jones and Bartlett, *Fundamentals of Fire Fighter Skills*, 1st Edition, 1st Printing, page 85.
    Answer: D

14. Reference: NFPA 1001, 6.2.2 and 6.2.2(A)(B)
    Delmar, *Firefighter's Handbook*, 2nd Edition, 1st Printing, page 54.
    IFSTA, *Essentials of Fire Fighting*, 4th Edition, 1st Printing, page 641.
    Jones and Bartlett, *Fundamentals of Fire Fighter Skills*, 1st Edition, 1st Printing, pages 77-78.
    Answer: A

15. Reference: NFPA 1001, 6.2.2 and 6.2.2(A)
    Delmar, *Firefighter's Handbook*, 2nd Edition, 1st Printing, page 49.
    IFSTA, *Essentials of Fire Fighting*, 4th Edition, 1st Printing, page 636.
    Jones and Bartlett, *Fundamentals of Fire Fighter Skills*, 1st Edition, 1st Printing, page 72.
    Answer: B

16. Reference: NFPA 1001, 6.3.1 and 6.3.1(A)(B)
Delmar, *Firefighter's Handbook*, 2nd Edition, 1st Printing, page 295.
IFSTA, *Essentials of Fire Fighting*, 4th Edition, 1st Printing, page 499.
Jones and Bartlett, *Fundamentals of Fire Fighter Skills*, 1st Edition, 1st Printing, page 525.
Answer: A

17. Reference: NFPA 1001, 6.3.1 and 6.3.1(A)(B)
Delmar, *Firefighter's Handbook*, 2nd Edition, 1st Printing, page 296.
IFSTA, *Essentials of Fire Fighting*, 4th Edition, 1st Printing, page 503.
Jones and Bartlett, *Fundamentals of Fire Fighter Skills*, 1st Edition, 1st Printing, page 526.
Answer: B

18. Reference: NFPA 1001, 6.3.1 and 6.3.1(A)(B)
Delmar, *Firefighter's Handbook*, 2nd Edition, 1st Printing, pages 297 and 601.
IFSTA, *Essentials of Fire Fighting*, 4th Edition, 1st Printing, page 499.
Jones and Bartlett, *Fundamentals of Fire Fighter Skills*, 1st Edition, 1st Printing, page 187.
Answer: C

19. Reference: NFPA 1001, 6.3.1 and 6.3.1(A)(B)
Delmar, *Firefighter's Handbook*, 2nd Edition, 1st Printing, page 297.
IFSTA, *Essentials of Fire Fighting*, 4th Edition, 1st Printing, pages 510-512.
Jones and Bartlett, *Fundamentals of Fire Fighter Skills*, 1st Edition, 1st Printing, page 527.
Answer: B

20. Reference: NFPA 1001, 6.3.1 and 6.3.1(A)(B)
Delmar, *Firefighter's Handbook*, 2nd Edition, 2nd Printing, page 295.
IFSTA, *Essentials of Fire Fighting*, 4th Edition, 1st Printing, page 499.
Jones and Bartlett, *Fundamentals of Fire Fighter Skills*, 1st Edition, 1st Printing, pages 527 and 529.
Answer: D

21. Reference: NFPA 1001, 6.3.1 and 6.3.1(A)(B)
Delmar, *Firefighter's Handbook*, 2nd Edition, 1st Printing, page 295.
IFSTA, *Essentials of Fire Fighting*, 4th Edition, 1st Printing, page 499.
Jones and Bartlett, *Fundamentals of Fire Fighter Skills*, 1st Edition, 1st Printing, page 529.
Answer: B

22. Reference: NFPA 1001, 6.3.1 and 6.3.1(A)(B)
Delmar, *Firefighter's Handbook*, 2nd Edition, 1st Printing, page 297.
IFSTA, *Essentials of Fire Fighting*, 4th Edition, 1st Printing, page 505.
Jones and Bartlett, *Fundamentals of Fire Fighter Skills*, 1st Edition, 1st Printing, page 525.
Answer: B

23. Reference: NFPA 1001, 6.3.1 and 6.3.1(A)(B)
Delmar, *Firefighter's Handbook*, 2nd Edition, 1st Printing, page 296.
IFSTA, *Essentials of Fire Fighting*, 4th Edition, 1st Printing, page 507.
Jones and Bartlett, *Fundamentals of Fire Fighter Skills*, 1st Edition, 1st Printing, page 526.
Answer: D

24. Reference: NFPA 1001, 6.3.1 and 6.3.1(A)(B)
Delmar, *Firefighter's Handbook*, 2nd Edition, 1st Printing, page 297.
IFSTA, *Essentials of Fire Fighting*, 4th Edition, 1st Printing, pages 58, 498-499.
Jones and Bartlett, *Fundamentals of Fire Fighter Skills*, 1st Edition, 1st Printing, page 526.
Answer: C

25. Reference: NFPA 1001, 6.3.1 and 6.3.1(A)(B)
Delmar, *Firefighter's Handbook*, 2nd Edition, 1st Printing, page 296.
Jones and Bartlett, *Fundamentals of Fire Fighter Skills*, 1st Edition, 1st Printing, page 526.
Answer: D

26. Reference: NFPA 1001, 6.3.1 and 6.3.1(A)(B)
Delmar, *Firefighter's Handbook*, 2nd Edition, 1st Printing, pages 297 and 601.
IFSTA, *Essentials of Fire Fighting*, 4th Edition, 1st Printing, page 58.
Jones and Bartlett, *Fundamentals of Fire Fighter Skills*, 1st Edition, 1st Printing, page 797.
Answer: A

27. Reference: NFPA 1001, 6.3.1 and 6.3.1(A)(B)
Delmar, *Firefighter's Handbook*, 2nd Edition, 1st Printing, page 626.
IFSTA, *Essentials of Fire Fighting*, 4th Edition, 1st Printing, page 530.
Jones and Bartlett, *Fundamentals of Fire Fighter Skills*, 1st Edition, 1st Printing, page 639.
Answer: A

28. Reference: NFPA 1001, 6.3.1 and 6.3.1(A)(B)
IFSTA, *Essentials of Fire Fighting*, 4th Edition, 1st Printing, page 530.
Jones and Bartlett, *Fundamentals of Fire Fighter Skills*, 1st Edition, 1st Printing, page 639.
Answer: B

29. Reference: NFPA 1001, 6.3.2 and 6.3.2(A)(B)
Delmar, *Firefighter's Handbook*, 2nd Edition, 1st Printing, page 612.
IFSTA, *Essentials of Fire Fighting*, 4th Edition, 1st Printing, page 538.
Jones and Bartlett, *Fundamentals of Fire Fighter Skills*, 1st Edition, 1st Printing, page 642.
Answer: D

30. Reference: NFPA 1001, 6.3.2 and 6.3.2(A)(B)
Delmar, *Firefighter's Handbook*, 2nd Edition, 1st Printing, page 353.
IFSTA, *Essentials of Fire Fighting*, 4th Edition, 1st Printing, page 65.
Jones and Bartlett, *Fundamentals of Fire Fighter Skills*, 1st Edition, 1st Printing, page 154.
Answer: A

31. Reference: NFPA 1001, 6.3.2 and 6.3.2(A)(B)
Delmar, *Firefighter's Handbook*, 2nd Edition, 1st Printing, page 361.
IFSTA, *Essentials of Fire Fighting*, 4th Edition, 1st Printing, pages 75-76.
Jones and Bartlett, *Fundamentals of Fire Fighter Skills*, 1st Edition, 1st Printing, page 427.
Answer: D

32. Reference: NFPA 1001, 6.3.2 and 6.3.2(A)(B)
Delmar, *Firefighter's Handbook*, 2nd Edition, 1st Printing, page 364.
IFSTA, *Essentials of Fire Fighting*, 4th Edition, 1st Printing, pages 73-74.
Jones and Bartlett, *Fundamentals of Fire Fighter Skills*, 1st Edition, 1st Printing, page 2.
Answer: D

33. Reference: NFPA 1001, 6.3.2 and 6.3.2(A)(B)
Delmar, *Firefighter's Handbook*, 2nd Edition, 1st Printing, page 520.
IFSTA, *Essentials of Fire Fighting*, 4th Edition, 1st Printing, pages 240-241.
Jones and Bartlett, *Fundamentals of Fire Fighter Skills*, 1st Edition, 1st Printing, page 299.
Answer: D

34. Reference: NFPA 1001, 6.3.2 and 6.3.2(A)(B)
Delmar, *Firefighter's Handbook*, 2nd Edition, 1st Printing, page 83.
IFSTA, *Essentials of Fire Fighting*, 4th Edition, 1st Printing, page 47.
Jones and Bartlett, *Fundamentals of Fire Fighter Skills*, 1st Edition, 1st Printing, page 127.
Answer: C

35. Reference: NFPA 1001, 6.3.2 and 6.3.2(A)(B)
Delmar, *Firefighter's Handbook*, 2nd Edition, 1st Printing, page 575.
IFSTA, *Essentials of Fire Fighting*, 4th Edition, 1st Printing, page 357.
Jones and Bartlett, *Fundamentals of Fire Fighter Skills*, 1st Edition, 1st Printing, page 414.
Answer: C

36. Reference: NFPA 1001, 6.3.2 and 6.3.2(A)(B)
Delmar, *Firefighter's Handbook*, 2nd Edition, 1st Printing, page 575.
IFSTA, *Essentials of Fire Fighting*, 4th Edition, 1st Printing, page 346.
Jones and Bartlett, *Fundamentals of Fire Fighter Skills*, 1st Edition, 1st Printing, page 423.
Answer: C

37. Reference: NFPA 1001, 6.3.2 and 6.3.2(A)(B)
Delmar, *Firefighter's Handbook*, 2nd Edition, 1st Printing, pages 575-576.
IFSTA, *Essentials of Fire Fighting*, 4th Edition, 1st Printing, page 354.
Jones and Bartlett, *Fundamentals of Fire Fighter Skills*, 1st Edition, 1st Printing, page 423.
Answer: B

38. Reference: NFPA 1001, 6.3.2 and 6.3.2(A)(B)
Delmar, *Firefighter's Handbook*, 2nd Edition, 1st Printing, page 575.
IFSTA, *Essentials of Fire Fighting*, 4th Edition, 1st Printing, pages 354 and 358.
Jones and Bartlett, *Fundamentals of Fire Fighter Skills*, 1st Edition, 1st Printing, page 429.
Answer: B

39. Reference: NFPA 1001, 6.3.2 and 6.3.2(A)(B)
Delmar, *Firefighter's Handbook*, 2nd Edition, 1st Printing, page 577.
IFSTA, *Essentials of Fire Fighting*, 4th Edition, 1st Printing, page 358.
Jones and Bartlett, *Fundamentals of Fire Fighter Skills*, 1st Edition, 1st Printing, page 431.
Answer: D

40. Reference: NFPA 1001, 6.3.2 and 6.3.2(A)(B)
Delmar, *Firefighter's Handbook*, 2nd Edition, 1st Printing, page 553.
IFSTA, *Essentials of Fire Fighting*, 4th Edition, 1st Printing, page 368.
Jones and Bartlett, *Fundamentals of Fire Fighter Skills*, 1st Edition, 1st Printing, page 420.
Answer: A

41. Reference: NFPA 1001, 6.3.2 and 6.3.2(A)(B)
Delmar, *Firefighter's Handbook*, 2nd Edition, 1st Printing, page 619.
IFSTA, *Essentials of Fire Fighting*, 4th Edition, 1st Printing, page 545.
Jones and Bartlett, *Fundamentals of Fire Fighter Skills*, 1st Edition, 1st Printing, page 437.
Answer: B

42. Reference: NFPA 1001, 6.3.2 and 6.3.2(A)(B)
Delmar, *Firefighter's Handbook*, 1st Edition, 2nd Printing, page 618.
IFSTA, *Essentials of Fire Fighting*, 4th Edition, 1st Printing, page 545.
Jones and Bartlett, *Fundamentals of Fire Fighter Skills*, 1st Edition, 1st Printing, page 518.
Answer: B

43. Reference: NFPA 1001, 6.3.2 and 6.3.2(A)(B)
Delmar, *Firefighter's Handbook*, 2nd Edition, 1st Printing, page 619.
IFSTA, *Essentials of Fire Fighting*, 4th Edition, 1st Printing, pages 545-546.
Jones and Bartlett, *Fundamentals of Fire Fighter Skills*, 1st Edition, 1st Printing, page 635.
Answer: D

44. Reference: NFPA 1001, 6.3.2 and 6.3.2(A)(B)
Delmar, *Firefighter's Handbook*, 2nd Edition, 1st Printing, page 565.
IFSTA, *Essentials of Fire Fighting*, 4th Edition, 1st Printing, page 488.
Jones and Bartlett, *Fundamentals of Fire Fighter Skills*, 1st Edition, 1st Printing, page 521.
Answer: C

45. Reference: NFPA 1001, 6.3.2 and 6.3.2(A)(B)
Delmar, *Firefighter's Handbook*, 2nd Edition, 1st Printing, page 354.
IFSTA, *Essentials of Fire Fighting*, 4th Edition, 1st Printing, page 66.
Jones and Bartlett, *Fundamentals of Fire Fighter Skills*, 1st Edition, 1st Printing, page 155.
Answer: C

46. Reference: NFPA 1001, 6.3.2 and 6.3.2(A)(B)
Delmar, *Firefighter's Handbook*, 2nd Edition, 1st Printing, page 205.
IFSTA, *Essentials of Fire Fighting*, 4th Edition, 1st Printing, page 391.
Jones and Bartlett, *Fundamentals of Fire Fighter Skills*, 1st Edition, 1st Printing, page 458.
Answer: C

47. Reference: NFPA 1001, 6.3.2 and 6.3.2(A)(B)
Delmar, *Firefighter's Handbook*, 2nd Edition, 1st Printing, pages 356-357.
IFSTA, *Essentials of Fire Fighting*, 4th Edition, 1st Printing, page 361.
Jones and Bartlett, *Fundamentals of Fire Fighter Skills*, 1st Edition, 1st Printing, page 158.
Answer: C

48. Reference: NFPA 1001, 6.3.2 and 6.3.2(A)(B)
Delmar, *Firefighter's Handbook*, 2nd Edition, 1st Printing, page 551.
IFSTA, *Essentials of Fire Fighting*, 4th Edition, 1st Printing, page 56.
Jones and Bartlett, *Fundamentals of Fire Fighter Skills*, 1st Edition, 1st Printing, page 128.
Answer: D

49. Reference: NFPA 1001, 6.3.2 and 6.3.2(A)(B)
Delmar, *Firefighter's Handbook*, 2nd Edition, 1st Printing, page 553.
IFSTA, *Essentials of Fire Fighting*, 4th Edition, 1st Printing, page 354.
Jones and Bartlett, *Fundamentals of Fire Fighter Skills*, 1st Edition, 1st Printing, page 411.
Answer: D

50. Reference: NFPA 1001, 6.3.2 and 6.3.2(A)(B)
Delmar, *Firefighter's Handbook*, 2nd Edition, 1st Printing, page 575.
IFSTA, *Essentials of Fire Fighting*, 4th Edition, 1st Printing, page 346.
Jones and Bartlett, *Fundamentals of Fire Fighter Skills*, 1st Edition, 1st Printing, page 423.
Answer: A

51. Reference: NFPA 1001, 6.3.2 and 6.3.2(A)(B)
Delmar, *Firefighter's Handbook*, 2nd Edition, 1st Printing, page 594.
IFSTA, *Essentials of Fire Fighting*, 4th Edition, 1st Printing, page 73.
Jones and Bartlett, *Fundamentals of Fire Fighter Skills*, 1st Edition, 1st Printing, page 290.
Answer: D

52. Reference: NFPA 1001, 6.3.2 and 6.3.2(A)(B)
Delmar, *Firefighter's Handbook*, 2nd Edition, 1st Printing, page 465.
IFSTA, *Essentials of Fire Fighting*, 4th Edition, 1st Printing, pages 176-178.
Jones and Bartlett, *Fundamentals of Fire Fighter Skills*, 1st Edition, 1st Printing, page 378.
Answer: A

53. Reference: NFPA 1001, 6.3.2 and 6.3.2(A)(B)
Delmar, *Firefighter's Handbook*, 2nd Edition, 1st Printing, page 579.
IFSTA, *Essentials of Fire Fighting*, 4th Edition, 1st Printing, pages 356 and 361.
Jones and Bartlett, *Fundamentals of Fire Fighter Skills*, 1st Edition, 1st Printing, pages 434-435.
Answer: A

54. Reference: NFPA 1001, 6.3.2 and 6.3.2(A)(B)
Delmar, *Firefighter's Handbook*, 2nd Edition, 1st Printing, page 88.
IFSTA, *Essentials of Fire Fighting*, 4th Edition, 1st Printing, page 534.
Jones and Bartlett, *Fundamentals of Fire Fighter Skills*, 1st Edition, 1st Printing, page 640.
Answer: C

55. Reference: NFPA 1001, 6.3.3 and 6.3.3(A)
Delmar, *Firefighter's Handbook*, 1st Edition, 2nd Printing, page 87.
IFSTA, *Essentials of Fire Fighting*, 4th Edition, 1st Printing, page 530.
Jones and Bartlett, *Fundamentals of Fire Fighter Skills*, 1st Edition, 1st Printing, page 640.
Answer: C

56. Reference: NFPA 1001, 6.3.3 and 6.3.3(A)(B)
Delmar, *Firefighter's Handbook*, 2nd Edition, 1st Printing, page 866.
IFSTA, *Essentials of Fire Fighting*, 4th Edition, 1st Printing, page 530.
Jones and Bartlett, *Fundamentals of Fire Fighter Skills*, 1st Edition, 1st Printing, page 641.
Answer: B

57. Reference: NFPA 1001, 6.3.3 and 6.3.3(A)(B)
IFSTA, *Essentials of Fire Fighting*, 4th Edition, 1st Printing, page 534.
Jones and Bartlett, *Fundamentals of Fire Fighter Skills*, 1st Edition, 1st Printing, page 638.
Answer: B

58. Reference: NFPA 1001, 6.3.4 and 6.3.4(A)(B)
Delmar, *Firefighter's Handbook*, 2nd Edition, 1st Printing, page 652.
IFSTA, *Essentials of Fire Fighting*, 4th Edition, 1st Printing, page 622.
Jones and Bartlett, *Fundamentals of Fire Fighter Skills*, 1st Edition, 1st Printing, page 972.
Answer: B

59. Reference: NFPA 1001, 6.3.4 and 6.3.4(A)(B)
Delmar, *Firefighter's Handbook*, 1st Edition, 1st Printing, page 653.
IFSTA, *Essentials of Fire Fighting*, 4th Edition, 1st Printing, page 629.
Jones and Bartlett, *Fundamentals of Fire Fighter Skills*, 1st Edition, 1st Printing, page 967.
Answer: B

60. Reference: NFPA 1001, 6.3.4 and 6.3.4(A)(B)
Delmar, *Firefighter's Handbook*, 2nd Edition, 1st Printing, page 653.
IFSTA, *Essentials of Fire Fighting*, 4th Edition, 1st Printing, page 629.
Jones and Bartlett, *Fundamentals of Fire Fighter Skills*, 1st Edition, 1st Printing, page 967.
Answer: D

61. Reference: NFPA 1001, 6.3.4 and 6.3.4(A)(B)
Delmar, *Firefighter's Handbook*, 2nd Edition, 1st Printing, page 650.
IFSTA, *Essentials of Fire Fighting*, 4th Edition, 1st Printing, page 597.
Jones and Bartlett, *Fundamentals of Fire Fighter Skills*, 1st Edition, 1st Printing, page 576.
Answer: D

62. Reference: NFPA 1001, 6.3.4 and 6.3.4(A)(B)
Delmar, *Firefighter's Handbook*, 1st Edition, 1st Printing, page 649.
IFSTA, *Essentials of Fire Fighting*, 4th Edition, 1st Printing, page 627.
Jones and Bartlett, *Fundamentals of Fire Fighter Skills*, 1st Edition, 1st Printing, page 574.
Answer: D

63. Reference: NFPA 1001, 6.4.1 and 6.4.1(A)(B)
Delmar, *Firefighter's Handbook*, 2nd Edition, 1st Printing, page 487.
IFSTA, *Essentials of Fire Fighting*, 4th Edition, 1st Printing, pages 198-199.
Jones and Bartlett, *Fundamentals of Fire Fighter Skills*, 1st Edition, 1st Printing, page 738.
Answer: A

64. Reference: NFPA 1001, 6.4.1 and 6.4.1(A)(B)
Delmar, *Firefighter's Handbook*, 2nd Edition, 1st Printing, pages 487-488.
IFSTA, *Essentials of Fire Fighting*, 4th Edition, 1st Printing, page 192.
Jones and Bartlett, *Fundamentals of Fire Fighter Skills*, 1st Edition, 1st Printing, page 738.
Answer: D

65. Reference: NFPA 1001, 6.4.1, 6.4.1(A)(B), 6.4.2, and 6.4.2(A)(B)
Delmar, *Firefighter's Handbook*, 2nd Edition, 1st Printing, page 478.
IFSTA, *Essentials of Fire Fighting*, 4th Edition, 1st Printing, page 185.
Jones and Bartlett, *Fundamentals of Fire Fighter Skills*, 1st Edition, 1st Printing, page 396.
Answer: C

66. Reference: NFPA 1001, 6.4.1 and 6.4.1(A)(B)
Delmar, *Firefighter's Handbook*, 2nd Edition, 1st Printing, page 488.
IFSTA, *Essentials of Fire Fighting*, 4th Edition, 1st Printing, page 202.
Jones and Bartlett, *Fundamentals of Fire Fighter Skills*, 1st Edition, 1st Printing, page 746.
Answer: B

67. Reference: NFPA 1001, 6.4.2 and 6.4.2(A)(B)
Delmar, *Firefighter's Handbook*, 2nd Edition, 1st Printing, page 496.
IFSTA, *Essentials of Fire Fighting*, 4th Edition, 1st Printing, page 208.
Jones and Bartlett, *Fundamentals of Fire Fighter Skills*, 1st Edition, 1st Printing, page 765.
Answer: A

68. Reference: NFPA 1001, 6.4.2 and 6.4.2(A)(B)
Delmar, *Firefighter's Handbook*, 2nd Edition, 1st Printing, page 497.
IFSTA, *Essentials of Fire Fighting*, 4th Edition, 1st Printing, pages 87, 92, and 549.
Jones and Bartlett, *Fundamentals of Fire Fighter Skills*, 1st Edition, 1st Printing, page 762.
Answer: A

69. Reference: NFPA 1001, 6.4.2 and 6.4.2(A)(B)
Delmar, *Firefighter's Handbook*, 2nd Edition, 1st Printing, page 467.
IFSTA, *Essentials of Fire Fighting*, 4th Edition, 1st Printing, page 184.
Jones and Bartlett, *Fundamentals of Fire Fighter Skills*, 1st Edition, 1st Printing, page 28.
Answer: B

70. Reference: NFPA 1001, 6.4.2 and 6.4.2(A)(B)
Delmar, *Firefighter's Handbook*, 2nd Edition, 1st Printing, page 469.
IFSTA, *Essentials of Fire Fighting*, 4th Edition, 1st Printing, pages 228-229.
Jones and Bartlett, *Fundamentals of Fire Fighter Skills*, 1st Edition, 1st Printing, page 392.
Answer: B

71. Reference: NFPA 1001, 6.4.2 and 6.4.2(A)(B)
Delmar, *Firefighter's Handbook*, 2nd Edition, 1st Printing, page 492.
IFSTA, *Essentials of Fire Fighting*, 4th Edition, 1st Printing, page 212.
Jones and Bartlett, *Fundamentals of Fire Fighter Skills*, 1st Edition, 1st Printing, pages 767-768.
Answer: A

72. Reference: NFPA 1001, 6.4.2 and 6.4.2(A)(B)
Delmar, *Firefighter's Handbook*, 2nd Edition, 1st Printing, pages 732-733.
IFSTA, *Essentials of Fire Fighting*, 4th Edition, 1st Printing, page 544.
Jones and Bartlett, *Fundamentals of Fire Fighter Skills*, 1st Edition, 1st Printing, page 27.
Answer: C

73. Reference: NFPA 1001, 6.5.1 and 6.5.1(A)(B)
Delmar, *Firefighter's Handbook*, 2nd Edition, 1st Printing, page 57.
IFSTA, *Essentials of Fire Fighting*, 4th Edition, 1st Printing, page 570.
Jones and Bartlett, *Fundamentals of Fire Fighter Skills*, 1st Edition, 1st Printing, page 927.
Answer: D

74. Reference: NFPA 1001, 6.5.1 and 6.5.1(A)(B)
Delmar, *Firefighter's Handbook*, 2nd Edition, 1st Printing, page 58.
IFSTA, *Essentials of Fire Fighting*, 4th Edition, 1st Printing, pages 637-638.
Jones and Bartlett, *Fundamentals of Fire Fighter Skills*, 1st Edition, 1st Printing, page 78.
Answer: B

75. Reference: NFPA 1001, 6.5.1 and 6.5.1(A)(B)
Delmar, *Firefighter's Handbook*, 2nd Edition, 1st Printing, page 309.
IFSTA, *Essentials of Fire Fighting*, 4th Edition, 1st Printing, page 570.
Jones and Bartlett, *Fundamentals of Fire Fighter Skills*, 1st Edition, 1st Printing, page 927.
Answer: D

76. Reference: NFPA 1001, 6.5.1 and 6.5.1(A)(B)
Delmar, *Firefighter's Handbook*, 2nd Edition, 1st Printing, page 310.
IFSTA, *Essentials of Fire Fighting*, 4th Edition, 1st Printing, page 560.
Jones and Bartlett, *Fundamentals of Fire Fighter Skills*, 1st Edition, 1st Printing, page 928.
Answer: B

77. Reference: NFPA 1001, 6.5.1 and 6.5.1(A)(B)
Delmar, *Firefighter's Handbook*, 2nd Edition, 1st Printing, page 311.
IFSTA, *Essentials of Fire Fighting*, 4th Edition, 1st Printing, page 563.
Jones and Bartlett, *Fundamentals of Fire Fighter Skills*, 1st Edition, 1st Printing, page 928.
Answer: C

78. Reference: NFPA 1001, 6.5.1 and 6.5.1(A)
Delmar, *Firefighter's Handbook*, 2nd Edition, 1st Printing, page 323.
IFSTA, *Essentials of Fire Fighting*, 4th Edition, 1st Printing, page 579.
Jones and Bartlett, *Fundamentals of Fire Fighter Skills*, 1st Edition, 1st Printing, page 942.
Answer: A

79. Reference: NFPA 1001, 6.5.1 and 6.5.1(A)
Delmar, *Firefighter's Handbook*, 2nd Edition, 1st Printing, page 319.
IFSTA, *Essentials of Fire Fighting*, 4th Edition, 1st Printing, page 580.
Jones and Bartlett, *Fundamentals of Fire Fighter Skills*, 1st Edition, 1st Printing, page 944.
Answer: B

80. Reference: NFPA 1001, 6.5.1 and 6.5.1(A)(B)
Delmar, *Firefighter's Handbook*, 2nd Edition, 1st Printing, page 324.
IFSTA, *Essentials of Fire Fighting*, 4th Edition, 1st Printing, pages 384 and 577.
Jones and Bartlett, *Fundamentals of Fire Fighter Skills*, 1st Edition, 1st Printing, page 940.
Answer: B

81. Reference: NFPA 1001, 6.5.1 and 6.5.1(A)(B)
Delmar, *Firefighter's Handbook*, 2nd Edition, 1st Printing, page 320.
IFSTA, *Essentials of Fire Fighting*, 4th Edition, 1st Printing, pages 581-582.
Jones and Bartlett, *Fundamentals of Fire Fighter Skills*, 1st Edition, 1st Printing, page 946.
Answer: C

82. Reference: NFPA 1001, 6.5.1 and 6.5.1(A)(B)
Delmar, *Firefighter's Handbook*, 2nd Edition, 1st Printing, page 322.
IFSTA, *Essentials of Fire Fighting*, 4th Edition, 1st Printing, pages 577-579.
Jones and Bartlett, *Fundamentals of Fire Fighter Skills*, 1st Edition, 1st Printing, page 942.
Answer: C

83. Reference: NFPA 1001, 6.5.1 and 6.5.1(A)(B)
Delmar, *Firefighter's Handbook*, 2nd Edition, 1st Printing, page 318.
IFSTA, *Essentials of Fire Fighting*, 4th Edition, 1st Printing, pages 577-579.
Jones and Bartlett, *Fundamentals of Fire Fighter Skills*, 1st Edition, 1st Printing, page 940.
Answer: D

84. Reference: NFPA 1001, 6.5.1 and 6.5.1(A)(B)
Delmar, *Firefighter's Handbook*, 2nd Edition, 1st Printing, pages 318 and 323.
IFSTA, *Essentials of Fire Fighting*, 4th Edition, 1st Printing, page 576.
Jones and Bartlett, *Fundamentals of Fire Fighter Skills*, 1st Edition, 1st Printing, page 940.
Answer: B

85. Reference: NFPA 1001, 6.5.1 and 6.5.1(A)(B)
Delmar, *Firefighter's Handbook*, 2nd Edition, 1st Printing, page 324.
IFSTA, *Essentials of Fire Fighting*, 4th Edition, 1st Printing, page 577.
Jones and Bartlett, *Fundamentals of Fire Fighter Skills*, 1st Edition, 1st Printing, page 940.
Answer: B

86. Reference: NFPA 1001, 6.5.1 and 6.5.1(A)
Delmar, *Firefighter's Handbook*, 2nd Edition, 1st Printing, page 324.
IFSTA, *Essentials of Fire Fighting*, 4th Edition, 1st Printing, page 577.
Jones and Bartlett, *Fundamentals of Fire Fighter Skills*, 1st Edition, 1st Printing, page 940.
Answer: D

87. Reference: NFPA 1001, 6.5.1 and 6.5.1(A)(B)
Delmar, *Firefighter's Handbook*, 2nd Edition, 1st Printing, page 324.
IFSTA, *Essentials of Fire Fighting*, 4th Edition, 1st Printing, page 577.
Jones and Bartlett, *Fundamentals of Fire Fighter Skills*, 1st Edition, 1st Printing, page 940.
Answer: B

88. Reference: NFPA 1001, 6.5.1 and 6.5.1(A)
Delmar, *Firefighter's Handbook*, 2nd Edition, 1st Printing, pages 318 and 323.
IFSTA, *Essentials of Fire Fighting*, 4th Edition, 1st Printing, pages 572-573.
Jones and Bartlett, *Fundamentals of Fire Fighter Skills*, 1st Edition, 1st Printing, page 939.
Answer: D

89. Reference: NFPA 1001, 6.5.1 and 6.5.1(A)(B)
Delmar, *Firefighter's Handbook*, 2nd Edition, 1st Printing, page 316.
IFSTA, *Essentials of Fire Fighting*, 4th Edition, 1st Printing, page 574.
Jones and Bartlett, *Fundamentals of Fire Fighter Skills*, 1st Edition, 1st Printing, page 936.
Answer: A

90. Reference: NFPA 1001, 6.5.1 and 6.5.1(A)(B)
Delmar, *Firefighter's Handbook*, 2nd Edition, 1st Printing, page 321.
IFSTA, *Essentials of Fire Fighting*, 4th Edition, 1st Printing, page 582.
Jones and Bartlett, *Fundamentals of Fire Fighter Skills*, 1st Edition, 1st Printing, page 947.
Answer: D

91. Reference: NFPA 1001, 6.5.1 and 6.5.1(A)(B)
Delmar, *Firefighter's Handbook*, 2nd Edition, 1st Printing, page 323.
IFSTA, *Essentials of Fire Fighting*, 4th Edition, 1st Printing, page 579.
Jones and Bartlett, *Fundamentals of Fire Fighter Skills*, 1st Edition, 1st Printing, pages 940-941.
Answer: B

92. Reference: NFPA 1001, 6.5.1 and 6.5.1(A)(B)
Delmar, *Firefighter's Handbook*, 2nd Edition, 1st Printing, page 314.
IFSTA, *Essentials of Fire Fighting*, 4th Edition, 1st Printing, page 572.
Jones and Bartlett, *Fundamentals of Fire Fighter Skills*, 1st Edition, 1st Printing, page 935.
Answer: C

93. Reference: NFPA 1001, 6.5.1 and 6.5.1(A)(B)
Delmar, *Firefighter's Handbook*, 2nd Edition, 1st Printing, pages 315 and 317.
IFSTA, *Essentials of Fire Fighting*, 4th Edition, 1st Printing, pages 574-575.
Jones and Bartlett, *Fundamentals of Fire Fighter Skills*, 1st Edition, 1st Printing, page 938.
Answer: D

94. Reference: NFPA 1001, 6.5.1 and 6.5.1(A)(B)
Delmar, *Firefighter's Handbook*, 2nd Edition, 1st Printing, page 318.
IFSTA, *Essentials of Fire Fighting*, 4th Edition, 1st Printing, page 581.
Jones and Bartlett, *Fundamentals of Fire Fighter Skills*, 1st Edition, 1st Printing, page 944.
Answer: C

95. Reference: NFPA 1001, 6.5.1 and 6.5.1(A)(B)
Delmar, *Firefighter's Handbook*, 2nd Edition, 1st Printing, page 318.
IFSTA, *Essentials of Fire Fighting*, 4th Edition, 1st Printing, page 581.
Jones and Bartlett, *Fundamentals of Fire Fighter Skills*, 1st Edition, 1st Printing, pages 944-945.
Answer: B

96. Reference: NFPA 1001, 6.5.1 and 6.5.1(A)(B)
Delmar, *Firefighter's Handbook*, 2nd Edition, 1st Printing, page 685.
IFSTA, *Essentials of Fire Fighting*, 4th Edition, 1st Printing, page 664.
Jones and Bartlett, *Fundamentals of Fire Fighter Skills*, 1st Edition, 1st Printing, page 651.
Answer: B

97. Reference: NFPA 1001, 6.5.1 and 6.5.1(A)
Delmar, *Firefighter's Handbook*, 2nd Edition, 1st Printing, page 685.
IFSTA, *Essentials of Fire Fighting*, 4th Edition, 1st Printing, page 664.
Jones and Bartlett, *Fundamentals of Fire Fighter Skills*, 1st Edition, 1st Printing, page 652.
Answer: A

98. Reference: NFPA 1001, 6.5.1 and 6.5.1(A)(B)
Delmar, *Firefighter's Handbook*, 2nd Edition, 1st Printing, pages 678-679.
IFSTA, *Essentials of Fire Fighting*, 4th Edition, 1st Printing, page 673.
Jones and Bartlett, *Fundamentals of Fire Fighter Skills*, 1st Edition, 1st Printing, page 923.
Answer: C

99. Reference: NFPA 1001, 6.5.1 and 6.5.1(A)(B)
Delmar, *Firefighter's Handbook*, 2nd Edition, 1st Printing, page 661.
IFSTA, *Essentials of Fire Fighting*, 4th Edition, 1st Printing, page 661.
Jones and Bartlett, *Fundamentals of Fire Fighter Skills*, 1st Edition, 1st Printing, page 651.
Answer: B

100. Reference: NFPA 1001, 6.5.1 and 6.5.1(A)(B)
Delmar, *Firefighter's Handbook*, 2nd Edition, 1st Printing, page 685.
IFSTA, *Essentials of Fire Fighting*, 4th Edition, 1st Printing, page 659.
Jones and Bartlett, *Fundamentals of Fire Fighter Skills*, 1st Edition, 1st Printing, page 650.
Answer: D

101. Reference: NFPA 1001, 6.5.1 and 6.5.1(A)(B)
Delmar, *Firefighter's Handbook*, 2nd Edition, 1st Printing, pages 311-314.
IFSTA, *Essentials of Fire Fighting*, 4th Edition, 1st Printing, pages 559-565.
Jones and Bartlett, *Fundamentals of Fire Fighter Skills*, 1st Edition, 1st Printing, pages 928-929.
Answer: A

102. Reference: NFPA 1001, 6.5.1 and 6.5.1(A)(B)
Delmar, *Firefighter's Handbook*, 2nd Edition, 1st Printing, page 311.
IFSTA, *Essentials of Fire Fighting*, 4th Edition, 1st Printing, page 564.
Jones and Bartlett, *Fundamentals of Fire Fighter Skills*, 1st Edition, 1st Printing, page 923.
Answer: D

103. Reference: NFPA 1001, 6.5.1 and 6.5.1(A)(B)
Delmar, *Firefighter's Handbook*, 2nd Edition, 1st Printing, page 685.
IFSTA, *Essentials of Fire Fighting*, 4th Edition, 1st Printing, page 656.
Jones and Bartlett, *Fundamentals of Fire Fighter Skills*, 1st Edition, 1st Printing, page 650.
Answer: D

104. Reference: NFPA 1001, 6.5.1 and 6.5.1(A)(B)
Delmar, *Firefighter's Handbook*, 2nd Edition, 1st Printing, page 56.
IFSTA, *Essentials of Fire Fighting*, 4th Edition, 1st Printing, page 559.
Jones and Bartlett, *Fundamentals of Fire Fighter Skills*, 1st Edition, 1st Printing, page 933.
Answer: C

105. Reference: NFPA 1001, 6.5.1 and 6.5.1(A)(B)
Delmar, *Firefighter's Handbook*, 2nd Edition, 1st Printing, page 312.
IFSTA, *Essentials of Fire Fighting*, 4th Edition, 1st Printing, page 563.
Jones and Bartlett, *Fundamentals of Fire Fighter Skills*, 1st Edition, 1st Printing, page 923.
Answer: A

106. Reference: NFPA 1001, 6.5.1 and 6.5.1(A)(B)
Delmar, *Firefighter's Handbook*, 2nd Edition, 1st Printing, page 310.
IFSTA, *Essentials of Fire Fighting*, 4th Edition, 1st Printing, page 562.
Jones and Bartlett, *Fundamentals of Fire Fighter Skills*, 1st Edition, 1st Printing, page 929.
Answer: D

107. Reference: NFPA 1001, 6.5.1 and 6.5.1(A)
Delmar, *Firefighter's Handbook*, 2nd Edition, 1st Printing, page 206.
IFSTA, *Essentials of Fire Fighting*, 4th Edition, 1st Printing, page 382.
Jones and Bartlett, *Fundamentals of Fire Fighter Skills*, 1st Edition, 1st Printing, page 448.
Answer: B

108. Reference: NFPA 1001, 6.5.1 and 6.5.1(A)
Delmar, *Firefighter's Handbook*, 2nd Edition, 1st Printing, page 204.
IFSTA, *Essentials of Fire Fighting*, 4th Edition, 1st Printing, page 382.
Jones and Bartlett, *Fundamentals of Fire Fighter Skills*, 1st Edition, 1st Printing, page 447.
Answer: B

109. Reference: NFPA 1001, 6.5.1 and 6.5.1(A)(B)
Delmar, *Firefighter's Handbook*, 2nd Edition, 1st Printing, page 205.
IFSTA, *Essentials of Fire Fighting*, 4th Edition, 1st Printing, page 380.
Jones and Bartlett, *Fundamentals of Fire Fighter Skills*, 1st Edition, 1st Printing, page 448.
Answer: C

110. Reference: NFPA 1001, 6.5.1 and 6.5.1(A)(B)
Delmar, *Firefighter's Handbook*, 2nd Edition, 1st Printing, page 205.
IFSTA, *Essentials of Fire Fighting*, 4th Edition, 1st Printing, pages 382-383.
Jones and Bartlett, *Fundamentals of Fire Fighter Skills*, 1st Edition, 1st Printing, page 448.
Answer: D

111. Reference: NFPA 1001, 6.5.1 and 6.5.1(A)(B)
Delmar, *Firefighter's Handbook*, 2nd Edition, 1st Printing, page 206.
IFSTA, *Essentials of Fire Fighting*, 4th Edition, 1st Printing, pages 382-383.
Jones and Bartlett, *Fundamentals of Fire Fighter Skills*, 1st Edition, 1st Printing, page 449.
Answer: D

112. Reference: NFPA 1001, 6.5.1 and 6.5.1(A)
Delmar, *Firefighter's Handbook*, 2nd Edition, 1st Printing, pages 214-216.
IFSTA, *Essentials of Fire Fighting*, 4th Edition, 1st Printing, pages 385-386.
Jones and Bartlett, *Fundamentals of Fire Fighter Skills*, 1st Edition, 1st Printing, page 454.
Answer: B

113. Reference: NFPA 1001, 6.5.1 and 6.5.1(A)
Delmar, *Firefighter's Handbook*, 2nd Edition, 1st Printing, pages 215-216.
IFSTA, *Essentials of Fire Fighting*, 4th Edition, 1st Printing, page 386.
Jones and Bartlett, *Fundamentals of Fire Fighter Skills*, 1st Edition, 1st Printing, page 455.
Answer: B

114. Reference: NFPA 1001, 6.5.1 and 6.5.1(A)
Delmar, *Firefighter's Handbook*, 2nd Edition, 1st Printing, pages 215-216.
IFSTA, *Essentials of Fire Fighting*, 4th Edition, 1st Printing, pages 386 and 389.
Jones and Bartlett, *Fundamentals of Fire Fighter Skills*, 1st Edition, 1st Printing, page 455.
Answer: A

115. Reference: NFPA 1001, 6.5.1 and 6.5.1(A)(B)
Delmar, *Firefighter's Handbook*, 2nd Edition, 1st Printing, page 209.
IFSTA, *Essentials of Fire Fighting*, 4th Edition, 1st Printing, page 384.
Jones and Bartlett, *Fundamentals of Fire Fighter Skills*, 1st Edition, 1st Printing, page 940.
Answer: D

116. Reference: NFPA 1001, 6.5.1 and 6.5.1(A)(B)
Delmar, *Firefighter's Handbook*, 2nd Edition, 1st Printing, page 208.
IFSTA, *Essentials of Fire Fighting*, 4th Edition, 1st Printing, page 387.
Jones and Bartlett, *Fundamentals of Fire Fighter Skills*, 1st Edition, 1st Printing, page 449.
Answer: C

117. Reference: NFPA 1001, 6.5.1 and 6.5.1(A)(B)
Delmar, *Firefighter's Handbook*, 2nd Edition, 1st Printing, page 205.
IFSTA, *Essentials of Fire Fighting*, 4th Edition, 1st Printing, pages 382-383.
Jones and Bartlett, *Fundamentals of Fire Fighter Skills*, 1st Edition, 1st Printing, page 448.
Answer: D

118. Reference: NFPA 1001, 6.5.1 and 6.5.1(A)(B)
Delmar, *Firefighter's Handbook*, 2nd Edition, 1st Printing, page 292.
IFSTA, *Essentials of Fire Fighting*, 4th Edition, 1st Printing, page 490.
Jones and Bartlett, *Fundamentals of Fire Fighter Skills*, 1st Edition, 1st Printing, page 466.
Answer: C

119. Reference: NFPA 1001, 6.5.1 and 6.5.1(A)(B)
Delmar, *Firefighter's Handbook*, 2nd Edition, 1st Printing, page 207.
IFSTA, *Essentials of Fire Fighting*, 4th Edition, 1st Printing, page 388.
Jones and Bartlett, *Fundamentals of Fire Fighter Skills*, 1st Edition, 1st Printing, page 450.
Answer: B

120. Reference: NFPA 1001, 6.5.1 and 6.5.1(A)(B)
Delmar, *Firefighter's Handbook*, 2nd Edition, 1st Printing, page 310.
IFSTA, *Essentials of Fire Fighting*, 4th Edition, 1st Printing, page 562.
Jones and Bartlett, *Fundamentals of Fire Fighter Skills*, 1st Edition, 1st Printing, page 929.
Answer: C

121. Reference: NFPA 1001, 6.5.1 and 6.5.1(A)
Delmar, *Firefighter's Handbook*, 2nd Edition, 1st Printing, page 56.
IFSTA, *Essentials of Fire Fighting*, 4th Edition, 1st Printing, page 562.
Jones and Bartlett, *Fundamentals of Fire Fighter Skills*, 1st Edition, 1st Printing, page 934.
Answer: D

122. Reference: NFPA 1001, 6.5.1 and 6.5.1(A)
Delmar, *Firefighter's Handbook*, 2nd Edition, 1st Printing, page 320.
IFSTA, *Essentials of Fire Fighting*, 4th Edition, 1st Printing, pages 581-582.
Jones and Bartlett, *Fundamentals of Fire Fighter Skills*, 1st Edition, 1st Printing, page 946.
Answer: C

123. Reference: NFPA 1001, 6.5.1 and 6.5.1(A)(B)
Delmar, *Firefighter's Handbook*, 2nd Edition, 1st Printing, page 321.
IFSTA, *Essentials of Fire Fighting*, 4th Edition, 1st Printing, pages 582-583.
Jones and Bartlett, *Fundamentals of Fire Fighter Skills*, 1st Edition, 1st Printing, page 947.
Answer: A

124. Reference: NFPA 1001, 6.5.1 and 6.5.1(A)(B)
Delmar, *Firefighter's Handbook*, 1st Edition, 1st Printing, pages 209-210.
IFSTA, *Essentials of Fire Fighting*, 4th Edition, 1st Printing, page 580.
Jones and Bartlett, *Fundamentals of Fire Fighter Skills*, 1st Edition, 1st Printing, page 940.
Answer: B

125. Reference: NFPA 1001, 6.5.1 and 6.5.1(A)(B)
Delmar, *Firefighter's Handbook*, 2nd Edition, 1st Printing, page 349.
IFSTA, *Essentials of Fire Fighting*, 4th Edition, 1st Printing, page 70.
Jones and Bartlett, *Fundamentals of Fire Fighter Skills*, 1st Edition, 1st Printing, page 150.
Answer: B

126. Reference: NFPA 1001, 6.5.1 and 6.5.1(A)
Delmar, *Firefighter's Handbook*, 2nd Edition, 1st Printing, page 354.
IFSTA, *Essentials of Fire Fighting*, 4th Edition, 1st Printing, page 66.
Jones and Bartlett, *Fundamentals of Fire Fighter Skills*, 1st Edition, 1st Printing, page 149.
Answer: A

127. Reference: NFPA 1001, 6.5.1 and 6.5.1(A)
Delmar, *Firefighter's Handbook*, 2nd Edition, 1st Printing, page 361.
IFSTA, *Essentials of Fire Fighting*, 4th Edition, 1st Printing, page 75.
Jones and Bartlett, *Fundamentals of Fire Fighter Skills*, 1st Edition, 1st Printing, page 427.
Answer: B

128. Reference: NFPA 1001, 6.5.1 and 6.5.1(A)
Delmar, *Firefighter's Handbook*, 2nd Edition, 1st Printing, page 659.
IFSTA, *Essentials of Fire Fighting*, 4th Edition, 1st Printing, page 655.
Jones and Bartlett, *Fundamentals of Fire Fighter Skills*, 1st Edition, 1st Printing, page 910.
Answer: A

129. Reference: NFPA 1001, 6.5.1 and 6.5.1(A)(B)
Delmar, *Firefighter's Handbook*, 2nd Edition, 1st Printing, page 685.
IFSTA, *Essentials of Fire Fighting*, 4th Edition, 1st Printing, pages 662-664.
Jones and Bartlett, *Fundamentals of Fire Fighter Skills*, 1st Edition, 1st Printing, page 651.
Answer: D

130. Reference: NFPA 1001, 6.5.1 and 6.5.1(A)(B)
Delmar, *Firefighter's Handbook*, 2nd Edition, 1st Printing, page 205.
IFSTA, *Essentials of Fire Fighting*, 4th Edition, 1st Printing, page 379.
Jones and Bartlett, *Fundamentals of Fire Fighter Skills*, 1st Edition, 1st Printing, pages 447-448.
Answer: D

131. Reference: NFPA 1001, 6.5.1 and 6.5.1(A)
Delmar, *Firefighter's Handbook*, 2nd Edition, 1st Printing, page 206.
IFSTA, *Essentials of Fire Fighting*, 4th Edition, 1st Printing, page 382.
Jones and Bartlett, *Fundamentals of Fire Fighter Skills*, 1st Edition, 1st Printing, page 449.
Answer: D

132. Reference: NFPA 1001, 6.5.1 and 6.5.1(A)
Delmar, *Firefighter's Handbook*, 2nd Edition, 1st Printing, page 206.
IFSTA, *Essentials of Fire Fighting*, 4th Edition, 1st Printing, page 383.
Jones and Bartlett, *Fundamentals of Fire Fighter Skills*, 1st Edition, 1st Printing, page 448.
Answer: C

133. Reference: NFPA 1001, 6.5.1 and 6.5.1(A)(B)
Delmar, *Firefighter's Handbook*, 2nd Edition, 1st Printing, page 310.
IFSTA, *Essentials of Fire Fighting*, 4th Edition, 1st Printing, pages 560-561.
Jones and Bartlett, *Fundamentals of Fire Fighter Skills*, 1st Edition, 1st Printing, page 929.
Answer: D

134. Reference: NFPA 1001, 6.5.1 and 6.5.1(A)
Delmar, *Firefighter's Handbook*, 2nd Edition, 1st Printing, page 314.
IFSTA, *Essentials of Fire Fighting*, 4th Edition, 1st Printing, pages 571-572.
Jones and Bartlett, *Fundamentals of Fire Fighter Skills*, 1st Edition, 1st Printing, page 935.
Answer: C

135. Reference: NFPA 1001, 6.5.1 and 6.5.1(A)(B)
Delmar, *Firefighter's Handbook*, 2nd Edition, 1st Printing, page 205.
IFSTA, *Essentials of Fire Fighting*, 4th Edition, 1st Printing, pages 382-383.
Jones and Bartlett, *Fundamentals of Fire Fighter Skills*, 1st Edition, 1st Printing, page 448.
Answer: A

136. Reference: NFPA 1001, 6.5.1 and 6.5.1(A)
Delmar, *Firefighter's Handbook*, 2nd Edition, 1st Printing, page 317.
IFSTA, *Essentials of Fire Fighting*, 4th Edition, 1st Printing, page 574.
Jones and Bartlett, *Fundamentals of Fire Fighter Skills*, 1st Edition, 1st Printing, page 938.
Answer: D

137. Reference: NFPA 1001, 6.5.1 and 6.5.1(A)
Delmar, *Firefighter's Handbook*, 1st Edition, 1st Printing, pages 319-321.
IFSTA, *Essentials of Fire Fighting*, 4th Edition, 1st Printing, pages 580-582.
Jones and Bartlett, *Fundamentals of Fire Fighter Skills*, 1st Edition, 1st Printing, page 943.
Answer: D

138. Reference: NFPA 1001, 6.5.1 and 6.5.1(A)
Delmar, *Firefighter's Handbook*, 2nd Edition, 1st Printing, page 318.
IFSTA, *Essentials of Fire Fighting*, 4th Edition, 1st Printing, page 580.
Jones and Bartlett, *Fundamentals of Fire Fighter Skills*, 1st Edition, 1st Printing, page 943.
Answer: A

139. Reference: NFPA 1001, 6.5.1 and 6.5.1(A)(B)
Delmar, *Firefighter's Handbook*, 2nd Edition, 1st Printing, page 329.
IFSTA, *Essentials of Fire Fighting*, 4th Edition, 1st Printing, pages 583-584.
Jones and Bartlett, *Fundamentals of Fire Fighter Skills*, 1st Edition, 1st Printing, page 946.
Answer: B

140. Reference: NFPA 1001, 6.5.1 and 6.5.1(A)(B)
Delmar, *Firefighter's Handbook*, 2nd Edition, 1st Printing, page 319.
IFSTA, *Essentials of Fire Fighting*, 4th Edition, 1st Printing, pages 580-581.
Jones and Bartlett, *Fundamentals of Fire Fighter Skills*, 1st Edition, 1st Printing, page 944.
Answer: D

141. Reference: NFPA 1001, 6.5.1 and 6.5.1(A)(B)
Delmar, *Firefighter's Handbook*, 2nd Edition, 1st Printing, page 325.
IFSTA, *Essentials of Fire Fighting*, 4th Edition, 1st Printing, page 577.
Jones and Bartlett, *Fundamentals of Fire Fighter Skills*, 1st Edition, 1st Printing, page 940.
Answer: C

142. Reference: NFPA 1001, 6.5.3 and 6.5.3(A)(B)
Delmar, *Firefighter's Handbook*, 2nd Edition, 1st Printing, page 274.
IFSTA, *Essentials of Fire Fighting*, 4th Edition, 1st Printing, pages 437 and 483.
Jones and Bartlett, *Fundamentals of Fire Fighter Skills*, 1st Edition, 1st Printing, page 479.
Answer: C

143. Reference: NFPA 1001, 6.5.4 and 6.5.4(A)(B)
Delmar, *Firefighter's Handbook*, 2nd Edition, 1st Printing, page 208.
IFSTA, *Essentials of Fire Fighting*, 4th Edition, 1st Printing, page 383.
Jones and Bartlett, *Fundamentals of Fire Fighter Skills*, 1st Edition, 1st Printing, page 449.
Answer: A

144. Reference: NFPA 1001, 6.5.4 and 6.5.4(A)(B)
Delmar, *Firefighter's Handbook*, 2nd Edition, 1st Printing, page 205.
IFSTA, *Essentials of Fire Fighting*, 4th Edition, 1st Printing, page 383.
Jones and Bartlett, *Fundamentals of Fire Fighter Skills*, 1st Edition, 1st Printing, page 448.
Answer: B

145. Reference: NFPA 1001, 6.5.4 and 6.5.4(A)(B)
Delmar, *Firefighter's Handbook*, 2nd Edition, 1st Printing, page 215.
IFSTA, *Essentials of Fire Fighting*, 4th Edition, 1st Printing, pages 386 and 389.
Jones and Bartlett, *Fundamentals of Fire Fighter Skills*, 1st Edition, 1st Printing, page 455.
Answer: D

146. Reference: NFPA 1001, 6.5.4 and 6.5.4(A)(B)
Delmar, *Firefighter's Handbook*, 2nd Edition, 1st Printing, page 215.
IFSTA, *Essentials of Fire Fighting*, 4th Edition, 1st Printing, page 389.
Jones and Bartlett, *Fundamentals of Fire Fighter Skills*, 1st Edition, 1st Printing, page 455.
Answer: B

147. Reference: NFPA 1001, 6.5.4 and 6.5.4(A)(B)
Delmar, *Firefighter's Handbook*, 2nd Edition, 1st Printing, page 215.
IFSTA, *Essentials of Fire Fighting*, 4th Edition, 1st Printing, pages 385-386.
Jones and Bartlett, *Fundamentals of Fire Fighter Skills*, 1st Edition, 1st Printing, pages 454-455.
Answer: D

148. Reference: NFPA 1001, 6.5.4 and 6.5.4(A)(B)
Delmar, *Firefighter's Handbook*, 2nd Edition, 1st Printing, page 215.
IFSTA, *Essentials of Fire Fighting*, 4th Edition, 1st Printing, page 394.
Jones and Bartlett, *Fundamentals of Fire Fighter Skills*, 1st Edition, 1st Printing, page 455.
Answer: B

149. Reference: NFPA 1001, 5.3.15 and 5.3.15(A)(B)
Delmar, *Firefighter's Handbook*, 2nd Edition, 1st Printing, page 208.
IFSTA, *Essentials of Fire Fighting*, 4th Edition, 1st Printing, page 390.
Jones and Bartlett, *Fundamentals of Fire Fighter Skills*, 1st Edition, 1st Printing, page 458.
Answer: B

150. Reference: NFPA 1001, 6.5.4 and 6.5.4(A)(B)
Delmar, *Firefighter's Handbook*, 2nd Edition, 1st Printing, page 207.
IFSTA, *Essentials of Fire Fighting*, 4th Edition, 1st Printing, page 387.
Jones and Bartlett, *Fundamentals of Fire Fighter Skills*, 1st Edition, 1st Printing, page 450.
Answer: D

BIBLIOGRAPHY FOR EXAM PREP: FIRE FIGHTER I AND II

1. NFPA, *Standard for Fire Fighter Professional Qualifications*, NFPA 1001, 2002
2. Delmar, *Firefighter's Handbook, Essentials of Firefighting and Emergency Response*, Second Edition
3. Delmar, *Firefighter's Handbook, Basic Essentials of Firefighting*, Second Edition
4. IFSTA, *Essentials of Fire Fighting*, Fourth Edition
5. Jones and Bartlett, *Fundamentals of Fire Fighter Skills*, First Edition

Performance Training Systems, Inc.
Training and testing that are on target!

Online examinations for the Fire and Emergency Medical Services

Registration

FREE OFFER - 150 ITEM PRACTICE TEST - VALUED AT $39.00

Complete registration form and fax it to (561) 863-1386.

_____
Name

_____
Title

_____
Department

_____
Address: Street

_____
City   State   Zip Code

_____
Telephone   Fax

_____
E-mail

Choose the tests that apply to your needs.

☐ Firefighter 1   ☐ Firefighter 2   ☐ Pumper Driver   ☐ Aerial Operator
☐ Fire Officer 1   ☐ Fire Officer 2   ☐ Fire Inspector 1   ☐ Fire Inspector 2
☐ Fire Instructor 1   ☐ Fire Instructor 2   ☐ HazMat Awareness   ☐ HazMat Operations
☐ HazMat Technician   ☐ Fire Investigator   ☐ EMT Basic

Signature:_____

Copyright 2000 Performance Training Systems, Inc.

Performance Training Systems, Inc.   International Association of Fire Chiefs

The only preparatory manuals to reference current NFPA standards!

The EXAM PREP Series

Each Exam Prep manual includes:
- Practice examinations
- Self-scoring guide with page references to multiple textbooks for further study
- Winning test-taking tips and helpful hints
- Coverage of the appropriate and current NFPA Standard

Titles in the Exam Prep series include:

Exam Prep: Airport Fire Fighter	ISBN: 0-7637-3764-X	$39.95
Exam Prep: Fire & Life Safety Educator I & II	ISBN: 0-7637-2854-3	$39.95
Exam Prep: Fire Apparatus Driver Operator	ISBN: 0-7637-2845-4	$39.95
Exam Prep: Fire Department Safety Officer	ISBN: 0-7637-2846-2	$39.95
Exam Prep: Fire Fighter I & II	ISBN: 0-7637-2847-0	$39.95
Exam Prep: Fire Inspector I & II	ISBN: 0-7637-2848-9	$39.95
Exam Prep: Fire Instructor I & II	ISBN: 0-7637-2762-8	$39.95
Exam Prep: Fire Investigator	ISBN: 0-7637-2849-7	$39.95
Exam Prep: Fire Officer I & II	ISBN: 0-7637-2761-X	$39.95
Exam Prep: Hazardous Materials Awareness & Operations	ISBN: 0-7637-2853-5	$29.95
Exam Prep: Hazardous Materials Technician	ISBN: 0-7637-2852-7	$39.95
Exam Prep: Technical Rescue—Structural Collapse and Confined Space	ISBN: 0-7637-2906-X	$39.95
Exam Prep: Technical Rescue—Vehicle/Machinery and Water/Ice	ISBN: 0-7637-2851-9	$39.95
Exam Prep: Technical Rescue—Ropes and Rigging	ISBN: 0-7637-2850-0	$39.95
Exam Prep: Telecommunicator I & II	ISBN: 0-7637-2856-X	$29.95
Exam Prep: Wildland Fire Fighter I & II	ISBN: 0-7637-2855-1	$29.95

Save 20% when you order 5 or more copies of any Exam Prep manual!

Don't take chances with other materials. Order your Exam Prep manuals today!

Yes! Please send me the following titles (specify quantity):

ISBN	Title:	Qty.:	Price:

Please include $6.00 shipping and handling for the first book and $1.00 for each additional book. For orders outside of the U.S., call 1-978-443-5000. CA, FL, MA, NY, SC, and TX customers, please add applicable sales tax.

❑ Check Enclosed payable to Jones & Bartlett Publishers
❑ Charge my: ❑ Mastercard   ❑ Visa   ❑ American Express   ❑ Discover

Card Number: _____   Expiration Date: _____
Signature: _____
Telephone: _____
Total: _____

Five ways to order!

Call: 1-800-832-0034
Fax: 1-978-443-8000
Email: info@jbpub.com
Visit: http://www.jbpub.com

Or mail in this completed order form:

Jones and Bartlett Publishers
40 Tall Pine Drive
Sudbury, MA 01776

Your order is risk free! If you are not completely satisfied with your purchase, return it within 30 days for a replacement copy or full refund.